Kleve Canteral
Manuella Raiol
Larissa Soares

Socio-environmental diagnosis: A case study of the municipality of Igarapé-Aç

Kleve Canteral
Manuella Raiol
Larissa Soares

Socio-environmental diagnosis: A case study of the municipality of Igarapé-Aç

ScienciaScripts

Imprint

Any brand names and product names mentioned in this book are subject to trademark, brand or patent protection and are trademarks or registered trademarks of their respective holders. The use of brand names, product names, common names, trade names, product descriptions etc. even without a particular marking in this work is in no way to be construed to mean that such names may be regarded as unrestricted in respect of trademark and brand protection legislation and could thus be used by anyone.

Cover image: www.ingimage.com

This book is a translation from the original published under ISBN 978-613-9-66272-2.

Publisher:
Sciencia Scripts
is a trademark of
Dodo Books Indian Ocean Ltd. and OmniScriptum S.R.L publishing group

120 High Road, East Finchley, London, N2 9ED, United Kingdom
Str. Armeneasca 28/1, office 1, Chisinau MD-2012, Republic of Moldova, Europe
Printed at: see last page
ISBN: 978-620-7-99430-4

SUMMARY

CHAPTER 1

INTRODUCTION

1.1. Presentation

The Amazon region is rich in terms of biodiversity, both plant and animal. Therefore, knowing how to manage this biological diversity is fundamental to sustaining it (MMA, 2002). This region is home to the municipality of Igarapé-Açu in the north-east of Pará, the target of this study. According to the Brazilian Institute of Geography and Statistics (IBGE), the municipality has a territorial extension of 786 km² and a population of 35,887 inhabitants, 63.82% of whom live in the urban area, according to the 2010 census.

According to Baena, Falesi and Dutra (1998), colonisation in the Amazon began in the north-eastern region of Pará with a strong exploration of plant species and agriculture, along with the introduction of crops such as black pepper, palm oil, coconut, among others, and cattle and buffalo breeding. The settlement of settlers in the region was favoured by the construction of the Bragança Railway in 1883 (LEANDRO and SILVA, 2012).

According to Silva et al. (2012), **Igarapé-Açu's main watercourses** are the Caripi and Maracanã rivers, and the Cumaru and São João streams, which are tributaries of the Maracanã and Jambuaçú rivers. Of these, the Maracanã River drains the hydrographic basin that defines the topographical boundaries of the municipality (VANZIN, 2014). In general, the most common types of soil in the region are: Latossolo amarelo textura média and concrencionário laterítico. The region also has Indiscriminate Hydromorphic and Alluvial soils (PORTELA; SOUZA, 2006).

The local vegetation is currently characterised by a mosaic of pastures, secondary vegetation, remnant forests and agricultural fields, characteristic of major anthropogenic alteration compared to the original landscape of terra firme forests, igapós, flooded fields and várzea, which are present in small proportions in the municipality. This profile is seen not only in Igarapé-Açu, but throughout the Bragantina Region in the north-east of the state of Pará, which the municipality is part of (KATO et al., 2006).

On the other hand, alterations to the landscape also cause changes in local biodiversity. The monocultures present in Igarapé-Açu, such as oil palm, not only contribute to soil degradation, but also to a decrease in biodiversity, since the habitat of many species is altered (MCT, 2013). These changes favour the emergence of pests and diseases in the plantations, resulting from their homogeneity and the local climatic conditions, characterised by Pachêco and Bastos (2006) as hot and humid, with an annual rainfall distribution of over 2,000 mm (SILVA et al., 2012).

1.2. Location and Access

The study site was the Igarapé-Açu School Farm (FEIGA) of the Federal Rural University of Amazonia (UFRA), 122 kilometres from Belém, the capital of the state of Pará, including the surrounding areas. An area 2 km from the Fazenda Escola was analysed, with an extension of 1,500 metres, mostly made up of oil palm plantations, as well as a region of piçarra extraction, black pepper plantations, houses, regenerating vegetation, areas of burning, and also a river that intersects this entire extension, surrounded by Cilician forest, called the Igarapé-açu river. Access to the site is via PA-127, Avenida Magalhães Barata and Travessa Sete de Setembro, as shown on map 02 (Annex 2).

Other areas around the Fazenda Escola were also visited, including the industrial estate and plantation areas of Agroindustrial Palmasa S/A, a company specialising in the extraction of palm and palm kernel oil, and the Caripi I stream, as shown on map 01 (Appendix 2).

1.3. Objectives

This activity aims, in general, to present the physiographic aspects of the study area in the municipality of Igarapé-Açu, including: climate, vegetation, soils, hydrography, geomorphology, geology, biodiversity, as well as establishing the socio-economic relationship of the study site. More specifically, the aim was to confirm, through prior observations, the veracity of the information gathered before visiting the site.

The aim was also to identify environmental problems in the region and suggest possible proposals to improve the environmental situation.

1.4. Methodology

For the best performance, the activity was divided into three stages: pre-field, field and post-field.

During the pre-fieldwork stage, local information was gathered on biodiversity, soils, climate, vegetation, limnology and geology. Based on these initial references, two maps were drawn up, one to locate the area and the other to classify land use and occupation, using the QGIS 2.18 software, which were used in the remote sensing, geoprocessing and soils assessments.

The second stage lasted four days from 11 to 14 September 2017, with visits to the various study areas. At this stage, observations were made of land use and occupation according to the map drawn up in the pre-field phase, confirming or adjusting the information collected prior to the trip, surveying drone images and collecting GPS (Global Positioning System) points. A number of limnological analyses were also carried out, as well as topographical data and measurements on the Caripi I stream, in order to calculate its flow rate and, together with the other assessments, make inferences about the environment studied.

The trip also included a visit to the Palmasa S/A company, both the industrial and plantation sectors, in order to contribute to the knowledge of Amazonian agro-ecosystems and observe their benefits and harms to society and the environment.

The biodiversity assessment was carried out at UFRA's Igarapé-Açu School Farm, by capturing animals using traps and searching for insects present in the region. The post-field phase consisted of comparing and analysing the data collected prior to the trip with the information obtained in the field, adjusting the classification of the land use and occupation map, and reaching the appropriate conclusions in the other areas studied.

CHAPTER 2

PHYSIOGRAPHIC ASPECTS

2.1. Weather

According to Lopes, Souza and Ferreira (2013), information related to climate and precipitation has reached prominence in the technical-scientific environment, on a regional and global scale. This prominence generates an increase in research, projects and products with data compiled from surface meteorological stations and derived from orbital satellites, thus making up global climatology. One of the main parameters to be analysed in relation to climate is precipitation, which is related to temperature, relative humidity and wind (ALBUQUERQUE et al., 2010).

Lopes, Souza and Ferreira (2013) and Albuquerque (2010) comment that rainfall in the Amazon, including the state of Pará, has its rainfall distribution related to large-scale atmospheric patterns, due to its proximity to the equator, with the main influencers being the South Atlantic Convergence Zones (SACZ) and the Intertropical Convergence Zones (ITCZ).

According to Silva et al. (2009), due to the importance of the municipality in the economy of northeastern Pará, there are already several studies related to the climate of Igarapé-Açu, which is classified by Koppen as type Am, categorising a hot and humid climate, with average annual rainfall of 2000 mm, average temperature of 28°C and relative humidity of 85%. The influence of **the ITCZ on the** locality can thus be seen in its hydrological regime, where the wettest periods are from March to May and December to February, and the least intense from September to November and June to August (ALBUQUERQUE, 2010).

The different social sectors, such as agriculture, tourism and industry, depend on the climatological relations of a location to better suit their daily activities, thus maximising their productive potential (LOPES, SOUZA and FERREIRA, 2013).

2.2. Vegetation

The concept of vegetation refers to the set of plant species peculiar to a terrain, country or region, and can vary according to the characteristics and factors in which it is found. Thus, factors such as solar radiation, water and soil type are essential to the diversity of species in a given area, and are also fundamental to the growth and development of vegetation. In addition, factors such as altitude, latitude, atmospheric pressure and the way air masses act condition plant variability (EMBRAPA, 2008).

Studies by Kunz (2009) show that the municipality of Igarapé-Açu used to be predominantly covered with vegetation characterised as evergreen and hydrophilous forest, which has now been replaced by secondary forest and areas used for agriculture.

Almeida (2010) defines secondary vegetation as that formed as a result of anthropogenic alterations to the original vegetation. In this sense, Igarapé-Açu, and specifically the piçarra extraction area, has secondary vegetation.

In general, vegetation performs various ecosystem services, as Nobre and Nobre (2002) point out, stating that the removal of atmospheric carbon contributes to mitigating the growing emission of greenhouse gases into the atmosphere. It also helps with hydrological regulation, climate control and erosion prevention.

2.3. Soils

The soil in the Igarapé-Açu region is mostly characterised as medium-textured yellow latosols and lateritic concretionary soils in the terra firma, as well as undifferentiated hydromorphic soils and alluvial soils in the floodplains.

Yellow Latosols are mineral, non-hydromorphic soils, uniform in terms of colour, texture and structure. They are dystrophic, alkalic and poor in the amount of exchangeable bases (Ca^{4-1-} , Mg^{4-1-} , K^+), which are more concentrated in the surface horizons due to the higher levels of organic matter (OLIVEIRA NETO and SILVA, 2011).

According to the Brazilian Soil Classification System (SIBICS), a brassolic B horizon is a subsurface mineral horizon where a high degree of weathering is evident, as evidenced by the almost complete alteration of easily altered minerals, followed by intense desilicification, base leaching and residual concentration of sesquioxides, and/or 1:1 clay minerals and primary minerals resistant to weathering. It is made up of varying amounts of iron and aluminium oxides, quartz and other minerals resistant to weathering.

The predominant soil in the region is considered to be well-drained with a predominance of clay and very clay texture, low natural fertility and low assimilable phosphorus values. In most cases, there is cohesion in the subsurface horizons, which can limit the good development of the root system.

In addition, the municipality of Igarapé-Açu is home to Yellow Argisols, which are deep, mineral soils with good drainage and whose main characteristic is a marked increase in clay content from the superficial A horizon to the subsurface B horizon, defining the diagnostic B textural horizon (Bt), which may or may not be waxy (EMBRAPA, 2006).

2.4. Hydrography

According to the International Hydrographic Organisation, hydrography is defined as the branch of applied science that deals with the measurement and description of the physical characteristics of the navigable part of the earth's surface and the adjacent coastal zones.

Brazil is one of the countries with one of the largest hydrographic complexes in the world, with rivers of great length and depth, as well as holding around 8 per cent of all the fresh water available on the earth's surface, which means it has great potential for generating electricity (SILVA, 2008).

Water resources are essential for most living organisms and are also fundamental for the development of an agricultural crop, directly influencing its production.

The municipality of Igarapé-Açu is made up of a network of rivers that govern the water dynamics of this place. The Maracanã River, which serves as the boundary with two other municipalities (Santa Maria and

Nova Timboteua), is the recipient of the vast majority of the streams that are present in Igarapé-Açu's hydrographic network, It has a meandering course all the way along the boundary with Nova Timboteua, which can be understood as winding rivers with many bends, often found in humid areas covered by riparian forest (PORTELA, 2010).

Another river of great importance in the municipality of Igarapé-Açu is the Caripi river, which rises in its interior and runs northwards towards Maracanã. It is formed by the union of the Primeiro Caripi stream, which receives the Raposo stream on its left bank and the Pupuca stream on its right.

The Jambú-Açu river is also part of the municipality's hydrological dynamics, with its tributaries on the right bank of the Igarapé-Açu river, which in turn gave rise to the municipality's name, being a direct tributary of the Marapanim river, receiving streams such as Pau Cheiroso, Colono and Santa Rita (PORTELA, 2010).

2.5. Geomorphology

According to Florenzano (2008), geomorphology is the science that studies the various forms of relief, according to their genesis and composition, together with the weathering that acts on them. Pointing out the importance of this study, as the relief establishes the best practices to be developed in a location, as well as interfering in the type of vegetation, course of rivers and types of soil present in a region, in other words, the relief is one of the parameters that defines the vulnerability of the location and the most appropriate type of management.

Geomorphology has various objects of study, with morphology being the variable most applied to environmental issues, as its main role is the qualitative descriptive presentation of landforms, and it is indicated as the starting point for subsequent analyses (FLORENZANO, 2008).

According to the IBGE geomorphology manual (2009), the municipality of Igarapé- Açu lies within the area known as the Amazonian Low Plateau - the Bragantina Zone, formed by sediments of ancient and current Tertiary and Quaternary age, from the Barreiras group. According to Valente et al. (2011), this region is configured as having a flat to gently undulating surface, with the presence of Latossolos and Argissolos Amarelos, with secondary forest vegetation at different levels of succession.

According to Silva et al. (2013), the structuring of relief on river terraces favours the formation of narrow alluvial valleys, with periodic flooding due to rainfall and tidal flows. Among the particularities of terrace modelling, as indicated by IBGE (2009), is the presence of slope breaks in relation to the riverbed, modifying its course. Thus, these characteristics, present in the municipality studied, serve as a divider between the region's river basins (CORTEZ, TAGLIARINI and TANCREDI, 2000). The topographic elevations (altimetry) present in the relief of the region studied are shown on map 8 (Annex 2), with the lowest altimetry being 28 metres and the highest being 57.6 metres.

2.6. Geology

The geology of the region is characterised by Cenozoic units and can be distinguished into two well-defined periods: the Tertiary period, which is represented by the Pirabas Formation, the Barreiras Formation and the Post-Barreiras. And the Quaternary period, which covers most of the north-east of the state of Pará, consisting of gravels, sands and unconsolidated clays, distinguished by recent alluvial deposits (COSTA et al., 1991).

The Pirabas Formation, which dates back to the Tertiary period, is lithologically composed of limestones of varying composition, sometimes interspersed with sandstones and leaflets, whose deposition has been attributed to a marine palaeoenvironment of shallow, warm waters. The Barreiras Formation, on the other hand, is made up of Cenozoic sediments that occur in the form of cliffs or terraces, made up of sandstones, siltstones, claystones and conglomerates with nuanced colours, predominantly yellow and red (GÓES et al., 1990; COSTA et al., 1991).

Various authors suggest that the Barreiras Formation should be separated into three different facies from the base to the top: conglomeratic, interpreted as being deposited under subaerial conditions and in the form of gravitational flow of debris; sandy clay, related to the regime of lakes in the more distal portions and associated with flooding towards the interior of the continent; and sandy clay, deposited under similar conditions to the conglomeratic facies. The last formation is the Post-Barreiras, which encompasses the yellowish sediments below the layers of the Barreiras Formation, which normally represents the top of the region's cliffs (IGREJA et al., 1990).

The Quaternary period was also called the Post-Tertiary, which was modified years later by Gervais and renamed the Holocene Epoch. This period comprises the smallest part of the Cenozoic era, corresponding only to the last 2Ma (SUGUIO et al., 2005). This period has been divided into two epochs: the Pleistocene and the Holocene. The former lasted from 1.8 to 0.01 million years and its peculiar characteristic is based on the appearance of hominids, in some categories, such as Australopithecus. The Holocene represents the period after the last glaciation, and life forms are basically those that exist today, with a few exceptions of life forms extinguished by the action of man himself (SOUZA et al., 2005).

During this period, the cold climate continued into the ice age, with warm periods. It was at this time that the insect flora and fauna became essentially modern, but several types of large mammals, now extinct, survived until the Pleistocene ice age (IGREJA et al., 1990).

The alternation of glacial and interglacial phenomena associated with sea level fluctuations, characteristic of the Quaternary period, leads to marine regressions and transgressions, a fact that makes it easy to deduce that the features linked to coastal environments are ephemeral, given the constant transformations (SUGUIO et al., 2005).

Silva (2011) comments that river terraces - geological formations characteristic of the Quaternary period - are formed due to the accumulation of materials on the banks of rivers and the difference between their thicknesses indicates that their formation occurred in phases, with a predominance of erosion caused by melting ice, which is a reflection of rising sea levels. The number of morphogenetic processes and products is enormous, such as river systems with their respective sedimentary deposits.

2.7. Biodiversity

The biodiversity of a place or region represents much of its characteristics. The term biodiversity refers to biological diversity to designate the variety of life forms at all levels, from microorganisms to wild flora and fauna, as well as the human species (ALHO, 2012).

According to Barbieri (2010), biodiversity therefore represents the living content of the Earth as a whole: everything that lives in the mountains, forests and oceans. It is found at all levels and encompasses not only all species of plants, animals and microorganisms, but also the ecological processes and ecosystems to which they belong.

It can also be a strong indicator of the environmental conditions present there, and indirectly demonstrate significant developments and changes in an area, whether this change is anthropogenic or natural. Fauna and flora also play an extremely important role as bioindicators (NIEMI and MCDONALD, 2004). According to the same author, an indicator organism is capable of measuring the environment's responses to anthropogenic disturbances.

With the profile of the Bragantina Zone marked by slash-and-burn agriculture, the ecosystems that represent the municipality are made up of rare capoeiras, invaded by juquira (typical weeds) and small manioc and chilli plantations (TRINDADE et al., 2015).

One of the characteristics of the region's land use, the use of fire to prepare areas in traditional agriculture (slash and burn) directly affects a very abundant biodiversity community, represented by arthropods, which can be considered natural predators of agricultural pests, such as ants and lepidopteran and coleopteran larvae (SANTOS et al., 2007). The type of land preparation system for a new crop can determine the diversity and abundance of local insect species (SANTOS et al., 2007).

In addition to the well-known cattle ranches and a few other pasture-raised species typical of the region, Venturieri (2006) points out that another option for generating income is the management of native melipon fauna, especially among family farming communities. This alternative involves the use of biodiversity with species that are not so common, but which demonstrate the important relationship between man and nature that can be established through various existing interactions.

2.8. Socioeconomics

According to the IBGE (2000; 2010), this municipality is subdivided into a municipal centre, corresponding to a small town, with approximately 19,489 inhabitants in 2000, rising to 21,207 in 2010. It also has agricultural colonies linked to the town centre, which total 43 and had 12,911 inhabitants in 2000, rising to 14,680 in 2010.

The municipality of Igarapé-Açu belongs to a region of ancient colonisation. Its economy is basically focused on agriculture, which is extensive in nature, uses family labour and is developed on smallholdings, ranging from 25 (twenty-five) to 100 (one hundred) hectares (MIRANDA, 2005).

In this region, the most commonly found agricultural crops are manioc (Manihot esculenta), black pepper (Piper nigrum), passion fruit (Passiflora edulis) and beans (Vigna unguiculata), which are widely planted by farmers in this municipality (VIEIRA, 2007). This is due to the fact that Igarapé-Açu is a municipality with a traditionally agricultural economy. According to this author, in addition to the use of **Commercial agro-forestry** systems **(SAFs) and backyard gardens,** family farmers maintain other land use systems, such as: annual crops (75 per cent), secondary forests (50 per cent), perennial crops (31.3 per cent), small animals (25 per cent), livestock (18.8 per cent) and fish farming (6.3 per cent).

Falesi and Galeão (2004) also point out in their work that the agro-economic activities practised in the north-eastern region of the state that most aggravate externalities in relation to natural resources are: rice, maize, beans and soya; oil palm cultivation; pipericulture and fruit cultivation, as well as livestock farming.A striking factor for local income was the different socio-economic conditions of the migrants who came to this region (capital, family labour, technical knowledge and adaptation to local conditions), which led to the adoption of the most diverse, intensive and complex production systems, marking a new logic of land use, articulation with the market and exploitation of natural resources (SILVA, 2010).As Silva (2010) explains, the region's economy is also marked by the cultivation of oil palm, which was introduced to Igarapé-Açu by the Japanese in plots ranging from 25 to 350 ha. The expansion of the crop was boosted by favourable bioclimatic conditions and the establishment of the company Agroindustrial Palmeira da Amazônia S.A (PALMASA) with tax incentives from the Federal Government, through the Superintendence for the Development of the Amazon (SUDAM).

CHAPTER 3

ENVIRONMENTAL ANALYSIS

3.1. Sensing and Geoprocessing

In a field study, it is necessary to first analyse the area being assessed. To this end, two pre-field maps were produced (Appendix 1) to adapt the study according to the elements present in the delimitation. The team's location map made it possible to choose the areas to be visited as a priority, while the classification map helped subdivide the area according to soil cover. Going out into the field provides a real recognition of the area, as well as pertinent adjustments to the maps.

3.1.1. Map production

All the maps were produced with the aid of QGIS 2.18 **software**, using the Google raster image provided in the Geoprocessing course and the cartographic bases initially required were obtained from the spatial database of the Army's Geographic Services Directorate. All the material was refined to better suit the reality of the location, using mechanisms available in the **software**. Each map has specific features related to its use. The maps in Annex 1 show information about the areas before the technical visit, in which a possible classification of the areas was indicated. After going to the site, it was possible to determine some corrections. The maps in Appendix 2 show the sites visited and their characteristics with greater veracity (maps 01, 02 and 03). Map 04 corresponds to the location of the permanent preservation area (APP), deforested areas and exposed soils and map 05 classifies the aforementioned items. Map 06 shows a mosaic made up of images captured by drone in the area referring to the Uesugi Agricultural Company, indicated on map 02 as visited area 2, where it is possible to see in greater detail the different uses and land cover present in the company, as well as the conditions of the lake dammed there. Map 07 shows the new classification of the area, with information refined after the field visit. Map 08 shows the relief formation in relation to the region's altimetry, which includes different topographic levels. To give reliability to the classification made on map 06, the Kappa index was applied, comparing it with a visual classification of the division.

3.1.2. Environmental analysis of the study area

As described above, each team has a pre-established area of around 1500 metres, as shown on map 02 (Appendix 2). The space allocated to the team is subdivided into visited areas 1 and 2, as can be seen on map 03 (Appendix 2), classified and categorised in point format, which are generated by walking using GPS.

3.1.2.1. Area visited 1

In visited area 1, the route ran from vertex 3 to vertex 2, according to map 02, the coordinates of which are shown in table 1. Along this route, it was possible to observe a high degree of conformity between the mapped area and what exists in the locality. In order to provide a more detailed interpretation, it was decided to subdivide the area into 3 groups, each of which has specific characteristics regarding land use and cover, vegetation and man's relationship with the environment.

Table 1: Coordinates of the vertices in the analysed area.

Vertex	Latitude	Longitude
V2	1°7'51.235" S	47°37'36,933" W
V3	1°8'40,039" S	47°37'36,977 W

Source: Authors, 2017.

The first group was classified as an area for extracting gravel. An area of around 4.02 hectares, surrounded by a wire fence and vegetation with signs of regeneration. The area shows fractions with a high level of compaction, indicators of vehicles passing through to transport the extracted material, as well as columns with the appearance of excavation, reaching a height of around 2.13 metres (figure 1 (a)). In the fraction above the column, there is more disaggregated gravel, with the appearance of surface runoff in the form of furrows, where if the area is not properly managed, there could be a loss of material through erosion and the formation of gullies. The soil at the site is classified as Plintossolo pétrico, with high iron oxidation apparent in its surroundings, figure 1 (b). The vegetation present in the delimitation of the area is made up of small shrubs, followed by denser vegetation in the background, with characteristics of capoeira in an advanced stage of succession, figure 1 (c). Between the extraction columns, it is possible to see undergrowth developing, signs of possible plant restructuring.

Figure 1: Piçarra extraction area. (a) height of the column of material for extraction. (b) concretions characteristic of Plintossolo pétrico. (c) extraction area with surrounding vegetation.

The second group consists of the buildings along the road in the team's area (figure 2 (a)). Problems related to man's interaction with the environment were observed around these properties. One of the most eminent problems is the dumping of domestic effluent without proper treatment or disposal, figure 2 (b).

Residential sewage can cause illnesses related to drinking contaminated water, skin diseases and strong odours. Along the entire length of the road that intersects the study area, it was possible to notice the burning of areas near the homes, indicating that the area was being prepared for possible planting, although no form of production was found. It should be noted that these buildings are only present on the right bank of the road, at a distance of 160 metres from the Igarapé-Açu river.

Figure 2: Disposition of housing and irregular rubbish dumping in the area.

Source: Authors, 2017.

The type of vegetation found at the site was characterised using photographic images taken by the botanical identifier at the Federal Rural University of Amazonia, determining the predominance of species on the right-hand side of the road (table 2) and, similarly, on the left-hand side of the road (table 3).

Table 2: Species most prevalent on the right-hand side of the road.

Scientific name	Common name	Category
Allamanda cathartica	Alamandra	Climber/Omamental
Arrabidaea Chica	Arrabedeia	Medicinal
Heliconia rostrata	Heliconea	Shrub
Maximiliana maripa	Red Inajá	Palm tree native to Pará
Sida sp.	Mallow	Weed/Medicinal
Clitoria fairchildiana	Vane	RAD/ Timber
Casearea arborea Urb.	Pau de Picos	Arboreal
Pueraria phaseoloides	Tropical Kudzu	Invader
Sapindus saponaria L.	Savoury	Medicnal
Cecropia pachystachya	Embaúba	A pioneer in degraded areas
Tachigalia paniculata Aubl.	Black Tachi	Timber

Sources: Authors, 2017.

The third group comprises the area with externally preserved vegetation. Throughout the entire length of the road, it was possible to see a tenuous preservation on its left bank, which predominantly has grasses with a darker green colour and in some places a drier appearance. To a lesser extent, it was possible to see medium-sized, highly robust trees. The vegetation found at the site characterises the APP region of the Igarapé-Açu river, which passes through the region studied.

The predominant species on the two sides helped to distinguish the banks of the road, the right side being highly anthropised and the left side more preserved due to the riverbed.

Table 3: Species most prevalent on the left-hand side of the road.

Scientific name	Common name	Category
Abarema jupunba	Abarema	Ornamental

Sclerolobium paniculatum Vogel	White taxi	Ornamental/ Timber
Jacaranda copaia	Pará-Pará	Ornamental/ Timber
Bauhinia cheilantha	Mororó/Cow's foot	Leguminous/ Medicinal
Talisea esculenta	Bull's eye	Medicinal

Source: Authors, 2017.

At the end of the road, there is a body of water, outside the pre-established limits, but its APP cuts out a small area of the delimitation. Because of this, the point is shown on map 03, so that the percentage of the preserved area can be quantified. Based on the maps analysed and the visit to the study site, it can be concluded that the area around the tributary of the Igarapé-Açu River is preserved, as is its APP, unlike the right bank of the road.

This highlights the importance of studying and mapping the area of interest before going into the field, as it allows for a better fit with the reality of the location, giving greater credibility to the work to be carried out.

3.1.2.2. Area visited 2

Visited area 2 refers to the property on which the Uesugi Agricultural Company is located (**l°7'54.902" S and 47°37'57.376"** W). It can be seen on map 03 and in more detail on mp 06, a mosaic of images captured by drone, providing better resolution and detail.

The Uesugi Agricultural Company (figure 3) was founded in 1973 by partner and owner Nilson Norique, with an area of around 200 hectares for planting black pepper and oil plants. According to the owner's account, at the time the company was set up, there was no land surrounding the company and there was a river spring in the area.

Figura 3:   Entrance to the Uesugi Agricultural Company - Igarapé Açu.

Source: Authors, 2017.

According to the owner, the company operates as follows: initially, the soil is managed using the conventional method of harrowing, liming and correcting it with chemical fertilisers, and then planting takes place. However, it should be noted that according to the information he provided, there has never been a soil analysis to adjust the content of the fertilisers applied to their real needs.

Oil palm production currently occupies around 3 hectares of the area, with oil palm processing taking place in partnership with Palmasa S/A, an oil palm processing company based in Igarapé-Açu. The production of black pepper is entirely reserved for export, with around 5,000 trees planted throughout the area (figure 4).

Figura 4:   Chilli peppers drying in the sun on the grounds of the Uesugi Agricultural Company.

Source: Authors, 2017.

The area around the company is currently undergoing rapid development, with several houses, but basic sanitation conditions in the region are still incipient. During the interview, Mr Noriaque said that when the project was set up, the river looked good, but today it is dammed and has a high concentration of macrophytes, as can be seen in figure 5 and map 06. According to Júnior (2015), macrophytes can be used as bioindicators of the environmental quality of rivers and lakes, making it possible to assess possible changes occurring in the environment. Their proliferation accelerates the eutrophication process, increasing biomass and decreasing the available oxygen content, and this proliferation is related to the concentration of macronutrients such as phosphorus and nitrogen. Therefore, the development of this phytoplankton community may be related both to the run-off of products used in the company's agricultural production and to the dumping of waste water from the company and adjacent properties.

Figura 5:   Impoundment of the water body within the company's extension.

Source: Authors, 2017.

5.1.3.   Kappa Index

Dainese (2001) points to the Kappa statistic as an excellent method for assessing the coherence of the classification of thematic maps with terrestrial reality. Its main advantage is that it uses all the elements of the error matrix to calculate the Kappa coefficient. According to Panzoni and Almeida (1996), the accuracy of thematic maps is determined through error matrices, made up of rows and columns of equal numbers, representing the comparison between the classification, which can be visual or digital, and the ground truth as a reference.

To classify the quality of the thematic map analysed, a table created by Landis and Koch in 1977 (figure 6) is used, relating it to the value of the Kappa coefficient.

Figura 6:   Table relating to the quality of thematic map classification.

Kappa value	Quality of the thematic map
<0,00	Bad
0,00-0,20	Bad
0,20-0,40	Reasonable
0,40-0,60	Good
0,60-0,80	Very good
0,80 - 1,00	Excellent

Source: Dainese, 2001.

The Kappa index was applied to map 07, with the classification after visiting the study site. The coefficient was 0.97, giving it an excellent quality rating, which makes the data expressed in this study more reliable.

3.2 Geology and Limnology

The field practice on 12 September 2017 was carried out in a public space intended for bathing in the surrounding community, with the presence of a body of water commonly known as a stream, which in the region was called Caripi I (figure 7).

Figura 7:   Panoramic view of the Caripi I stream in Igarapé-Açu/PA.

Source: Authors, 2017.

7.2.1.   Characteristics of the catchment area and watercourse under study

The main watercourses in the municipality of Igarapé-Açu are the Caripi and Maracanã rivers, as well as the Jambu-Açu, Cumaru and São João streams, both of which are tributaries of the Maracanã River (SILVA, 2009). The Caripi I stream drains into the Caripi river basin, which predominates in much of the municipality of Igarapé-Açu, and is a tributary of the Maracanã river. The Caripi River basin lies to the north in the Salgado micro-region and to the south in the Bragantina micro-region (TAMASAUKA et al., 2017). According to Fapespa (2016), the drainage is generally dentritic to dense, given the area of sedimentary lithology it runs through.

Despite being identified as a type of igarapé, which are generally **small in size, this watercourse has considerable dimensions, and its channel is** classified as 3ª to 4ª order according to its depth. **The body of water has** lower and shallower areas along **its** length, so it can be classified as an asymmetrical channel. As it has curvilinear and uniform shapes throughout its structure, it is indicated as a meandering channel, as shown in figure 8. This type of channel is generally found in tropical regions.

Figura 8:   Aerial view of the Caripi I stream, highlighting the sinuosity of the channel obtained using a drone. Source: Authors, 2017.

In terms of fluvial geomorphology, its structure is made up of a channel and terrace, with the existing escarpment being disregarded due to the construction of the small bridge that intersects part of the stretch. At the study site, it could be seen that the flow of water is reduced due to a narrowing in the channel. Its waters are classified as black water due to their turbid appearance and the presence of organic matter.

1.1.2.   Uses and occupation

Siltation was identified along the watercourse and obstacles in its surroundings such as the remains of logs, probably due to the extensive vegetation that surrounds the creek and contributes to the deposition of organic matter through dry leaves and decaying wood. It was also noted that in addition to recreational bathing, the banks of the stream are used for washing clothes.

1.1.3.   Collection and evaluation of physical and chemical parameters

Physical and chemical water parameter values were collected using a HANNA HI 9828 multi-parameter probe, as shown in Table 4 below.

Table 4: Values of physical and chemical parameters measured in the Caripi I stream.

Parameters	Values
Temperature	26.08^0 C
pH	6,4
Oxygen	5.3 mg/l
Conductivity	18 µS/cm
Salinity	0,01
Dissolved Oxygen	64,8%
Total Dissolved Solids	9 ppm

Source: Authors, 2017.

The temperature of surface waters results mainly from the climatic conditions of the area, varying with depth and time of day, and directly affects many of the physical, chemical and biological characteristics of ecosystems (WAICHMAN, 2002). The temperature value of 26.08 °C is close to the values generally found in the literature, which vary between 28 °C and 30 °C for Class II bodies of water, which this igarapé falls into. Values above 33 °C can cause changes in chemical and biological reactions, favouring the development of microorganisms and intensifying tastes and odours (RIBEIRO et al., 2010).

In relation to the pH value, according to CONAMA resolution 357/2005, this parameter must vary between 6.0 and 9.0, with 6.4 being a value that is close to those mentioned, and it should be noted that the predominant pH values in waters in the state of Pará are generally low, characterising acidic waters.

Electrical conductivity is a parameter that expresses the capacity of water to conduct electrical current through dissolved ions. This parameter provides important information both on the metabolism of the aquatic ecosystem and on other biogeochemical processes taking place in the river basin, helping to detect possible sources of pollution (ESTEVES, 1998).

According to CONAMA 357/2005, salinity refers to the dissolved ions in the water and for Class II waters the ideal value is equal to or less than 0.5 per cent, with values of 0.01 per cent being found for this study.

As one of the most important parameters for the survival of most aquatic organisms, dissolved oxygen plays several roles in the hydrological environment, including the production of photosynthesis and participation in the stabilisation of organic matter through the self-depuration of the water body. According to Embrapa (2011), a dissolved oxygen saturation of over 60 per cent is acceptable for most aquatic species, while lower levels indicate oxygen-poor waters. In this case, a level of 64% was found.

The aforementioned legislation stipulates that total dissolved solids should not exceed 500 mg/L for Class II waters. This study found a value of 9 ppm or 9 mg/l, well below the stipulated maximum limits.

1.3. Meteorology and Hydrology

The hydrology subject's practical work on the water body under study consisted of measuring flow rates and velocity profiles using a windlass (figure 9). This is a device with a propeller at the end, which fits onto a 2 metre millimetre rod. It is connected to a counter of the turns produced by its propeller, which will indicate the total number of turns resulting from the water flow in a given time.

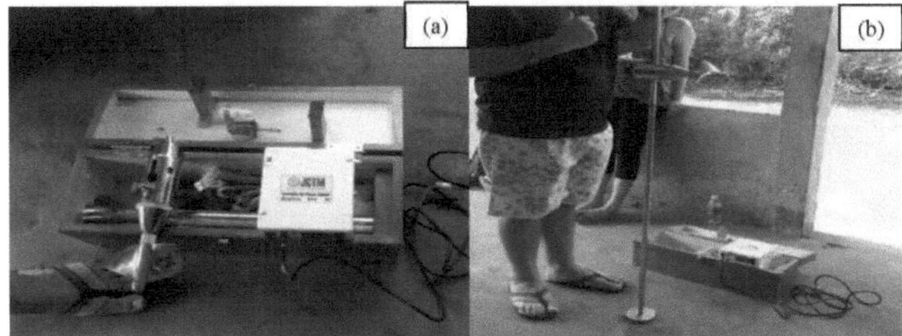

Figure 9:   Equipment case containing windlass and spin counter (a) and windlass fitted to a millimetre rod for use underwater (b).

Source: Authors, 2017.

First, the total width of the river profile was delimited using a tape measure, as shown in Figure 10, obtaining a value of 7.35 metres. The areas were then subdivided equally into 5 smaller sections along the measured width.

Figura 10:   Measuring the cross section of the Caripi I stream with a tape measure.

Source: Authors, 2017.

Points were also collected along the river profile (1 ° 7' 35.2" S and 47 ° 34' 20.291" W) using a RUIDE - RTS 822R3 Total Station to obtain the igarapé's altimetry. For a more precise analysis, Grapher 6.0 software was used, showing the depth of the studied section of the igarapé in a representative manner, figure 11. The study site is located at coordinates:

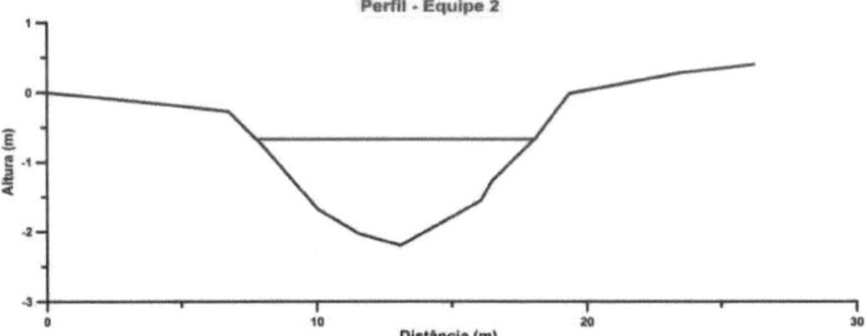

Figure 11:   Schematic drawing of the section of the Caripi I stream according to its height and depth.

Source: Authors, 2017.

At each section point, the depth was measured with the millimetre rod, and the windlass was fitted at a height of 60%, measuring from the surface downwards, as shown in Figure 12, and then switching on the spin counter to carry out the rest of the operation.

Figure 12: Use of the windlass equipment by the work team.

Source: Authors, 2017.

3.3.1   Results and Discussions on Flow Calculation

The results obtained and calculated can be seen in Table 5 below: Table 5: Characteristics of the Caripi 1 stream section, obtained by the windlass.

Verti ca is	Depth	N° Spins	H (Height)	Speed* (m/s)	Area (m)²	L (m)	Flow (m³ /s)
0	0	0	0	0	0	0	0
1	0,75	12	0,45	2,93	0,66	1,47	1,93
2	0,93	40	0,56	9,74	0,82	1,47	7,99
3	1,16	30	0,70	7,30	1,03	1,47	7,52
4	0,85	33	0,51	8,03	0,75	1,47	6,02
5	0,53	7	0,32	1,72	0,47	1,47	0,81

Source: Authors, 2017. *Obtained from the Windlass Equation: V (m/s): 0.01700114+0.24297688*(N° revolutions).

The presence of aquatic plants was identified in the section analysed in the field known as Victoria Regia (Amazon Victoria), these possibly reduce the flow of the current at the point where they are located, which in this case were vertical points 1 and 5, these points are also where the banks of the watercourse are located, thus obtaining shallower depths and presenting less river dynamics.

Obstacles such as rock fragments and logs were identified at vertical point 3, which explains the reduction in speed compared to the next point, which is shallower but faster.

In their study on the management of water resources, Oliveira and Canellas (2012) emphasise the need to forecast inflows to reservoirs in order to establish the availability of electricity generation at a plant. According to the same authors, this parameter also makes up the series of hydro-meteorological data that is used to size the most important equipment and structures.

Silveira and Souza (2012) mention that establishing the relationship between parameters such as rainfall and flow, in addition to characterising the behaviour of a river basin, provides important information for municipal management as it allows critical periods to be determined that are subject to catastrophes such as floods and mass movements.

1.4. Soils

1.4.1. Piçarra extraction area

The process of mineral extraction has become largely responsible for Brazil's economic growth, and its consequences are becoming increasingly noticeable. In the municipality of Igarapé-Açu, where the study was carried out, **the occurrence of Piçarra extraction** was identified in one of the **areas visited (l°8'32.705"S,47°37'42.571"W), as** shown in map 05 in Annex 2.

1.4.1.1. Soil classification and the formation processes involved

Based on observations and the type of activity carried out, it was found that the type of soil in this location is called Plintossolo pétrico, as illustrated in figure 13 (a). The total estimated volume of soil removed in the exploration was 62,310 m^3 and could be quantified based on on-site measurements, Figure 13 (b), and with the prior help of QGIS 2.18 software.

According to the Brazilian Soil Classification System (SIBICS), Plinthosols **"comprise water mineral soils subject to the temporary effect of excess moisture, and poorly drained", and are among the most common soils in the northern region of Brazil due to its** hot and humid **climate.** Petric plinthosols fall into the 2nd categorical level of SIBICS and are characterised by having a concretionary horizon or lithoplastic horizon. This type of soil is also influenced by the seasonal variation of the water table, which is generally responsible for the threshold of ferruginous concretions in soils (EMBRAPA, 1999; OLIVEIRA, 2001). In order to concretely determine the type of soil in the environment, it is necessary to carry out soil sampling and physical analyses in order to determine the texture and thus the class.

The genesis of Plinthosol is associated with the segregation, mobilisation, transport and concentration of iron (Fe) ions and compounds, which can come either from the source material or be translocated from other horizons or from adjacent higher areas (DRIESSEN & DUDAL, 1989). In addition, Plintite can be established as a formation composed of a mixture of clay material with particles of quartz and other minerals, poor in carbon and rich in Fe, or Fe and Al, irreversibly consolidated from numerous wetting and drying cycles.

Figure 13: (a) Measurement of the area. (b) Plinthosol of the site.

Source: Authors, 2017.

A morphological assessment of the soils was carried out at the study site in accordance with the Pedology Technical Manual, the information for which is contained in Annex 4.

1.4.1.2. Identification of impacts and vegetation in the extraction area

Piçarra extraction generates various environmental impacts in the area where it takes place, such as the removal of vegetation, excavations, de-characterisation of the terrain, modification of the local landscape, alteration of erosion processes and gullies (figure 14 (a)). Because of the location of the extraction site and the inefficiency of inspection, the exploitation points grow and with them the impacts expand and can influence the lives of the population living near the site. The effects of extraction can be identified mainly in relation to the soil, such as: loss of vegetation cover, which intensifies erosion and leaching processes, and also compaction, due to the fact that extraction is carried out with heavy machinery, which severely affects soil conditions and the possibilities for new species to emerge (FERREIRA et al., 2010).

Vegetation is the soil's natural protection against water erosion, so the loss of vegetation cover aggravates erosion processes and is also one of the main factors in soil impoverishment. Erosion is a slow and gradual natural process in which soil particles are detached, dragged and deposited by water and wind (NUNES et al., 2011). However, it can be intensified by anthropogenic actions, such as deforestation, agricultural activities and inadequate soil management, as occurred in the municipality visited. Some erosion points were identified in and around the region (1º 8'12.476"S,47º 37'38.712"W) as shown in figure 14 (b).

Figura 14:   (a) Gullies identified in the area. (b) Identification of erosion points.

Source: Authors, 2017.

In addition, the removal of gravel has a connection with soil compaction, due to the use of heavy machinery related to some of the soil's physical and mechanical properties, especially those with a higher mass/volume ratio, since compaction is an increase in mass or a reduction in the soil's pore space, for a given volume of the soil, so if it is not possible to prove soil compaction through visual analyses, quantitative analyses can be carried out, and the analyses of the physical properties most commonly presented in the literature can be adopted, which are indicative of the occurrence of problems: porosity, density, water infiltration and resistance to penetration.

Furthermore, research such as that carried out by Secco et al. (2004) has proven the relationship between soil compaction and the poor development of some plant species. Therefore, in order to really assess whether there is compaction and its level, it is extremely important to carry out physical analyses.

In addition to the anthropogenic factors that affect the piçarra removal area, which can hinder the emergence of vegetation, there are also natural factors related to the type of **soil. According to Moreira & Oliveira (2008), Plintossolos pétricos are "usually poor** in terms of natural fertility and, due to the impediment to mechanisation and root penetration, **represented by concretions". Thus, due to this lack of nutrients, it is common for** vegetation to not grow well in stony plinthosols.

Based on all the observations that have been made and mentioned above, both in relation to natural conditions and the impacts caused by anthropogenic actions, it is expected that the locality does not have a vast expanse of vegetation and biological diversity. However, it was observed that despite all the impacts suffered, the region shows a natural regeneration of vegetation with the presence of some plant species, shown in table 6 and figure 15.

Chart 6: Plant species in the natural regeneration area.

Scientific name	Common name	Category
Arrabidaea Chica	Arrabedeia	Medicinal
Vismia sp	Seal	Medicinal
Ravenala guianensis Peterson	Sororoca	Medicinal
Borreria verticillata (L.) G. Mey	Button broom	Medicinal
Mauritia Flexuosa	Buriti	Palm tree
Maximiliana regia	Inajazeiro	Palm tree

Nephrolepis exaltata	Fern	Ornamental
Cecropia pachystachya	Embaúba	A pioneer in degraded areas

Source: Authors

Analysing table 6, it can be seen that twelve different tree species were identified with the help of the UFRA botanical identifier. Four types of plants with medicinal potential, two palm trees, one ornamental plant and one pioneer species in degraded areas were observed.

The medicinal plants found in the area were Arrabedeia (Arrabidaea Chica), Lacre (Vismia sp), Sororoca (Ravenala guianensis Peterson) and Vassourinha de botão [Borreria verticillata (L.) G. Mey]. Amongst phytotherapeutic plants, there are many healing properties that can be found in the leaf, the stem and even the root, and the effects are varied, such as anti-inflammatory, healing, treatment against rheumatism, protection against infections in dermatitis and against diabetes (COSTA and LIMA, 1989).

A variety of palm trees were also found: Buriti (Mauritia Flexuosa) and Inajazeiro (Maximiliana regia). Buriti is a plant of great importance, as it is almost completely utilised, with a variety of uses, such as extracting oil, using the wood from the stem, using the pulp to make sweets, and especially for local handicrafts, among other uses given to the palm. The Inajazeiro is typical of dry places, and its use is mainly for the production of biodiesel from the pulp and kernels of the inajá (KELLER, 2014; MOTA and DE FRANÇA, 2007).

In addition, ferns (Cecropia pachystachya), which are ornamental plants, and Embaúba (Cecropia pachystachya) were found. Embaúba is considered a pioneer species that occurs in the Atlantic Forest, mainly at the edge of the forest or in secondary forests, characteristic of humid soils and riverbanks (GODOI and TAKAKI, 2005). However, the aforementioned characteristics are not the reality found in the piçarra extraction region, but the presence of this plant can be explained by the high temperature of the environment. According to Godoi and Takaki (2005), embaúba seeds germinate more efficiently when exposed to average temperatures of 30 °C, with variations between 25 and 35 °C. The species was therefore considered a pioneer in the recovery of the region.

Figura 15:   Vegetation in the Piçarra retreat area, figures a, b, c, d, e, f represent the different species found.

Source: Authors, 2017.

3.4.1.3.   Alternatives for recovering/regenerating the area

Given all the changes to the region's environmental characteristics due to mineral exploitation, large degraded areas appear at the end of exploitation, **which become "empty spaces", because as Kopezinski (2000) states, the "extracted mineral goods no longer return to the site, they remain in circulation, serving man and his needs".** However, public policies have emerged with the aim of forcing mining companies to recover the degraded area or give it another purpose, as shown in

Article 225, paragraph 2 of the Federal Constitution, dedicated to the environment and the obligation of those who exploit mineral resources to restore the degraded environment.

Article 225 states that,

25

> Everyone has the right to an ecologically balanced environment, which is a good for the common use of the people and essential to a healthy quality of life, imposing on the public authorities and the community the duty to defend and preserve it for present and future generations.
> § Paragraph 2 - Anyone who exploits mineral resources is obliged to restore the degraded environment, in accordance with the technical solution required by the competent public body, in accordance with the law.

This being the case, an alternative way of reclaiming the space and preventing loss due to water erosion is to use the reclamation methodology of the Brazilian Agricultural Research Corporation (EMBRAPA), which introduces plant species such as leguminous trees and shrubs capable of growing under adverse conditions, making a plant-rhizobium-mycorrhizal fungus association that provides rapid plant growth, regardless of the nitrogen available in the soil. It is also necessary to increase the content of organic matter and biological activity by adding plant material via leaf litter. An alternative to applying leaf litter is to add a layer of topsoil over the area to be planted.

According to Magalhães (2002), the discourse on sustainability and development is orientated towards the obligation to treat it in a multidimensional way that articulates economic, political, ethical, social, cultural and ecological aspects, avoiding the reductionism of the past. Another sustainable practice for rehabilitation is local tourism, and several cities have already adopted this recovery measure, such as Curitiba, which has set up a park where in the past it was a mineral exploration area. Thus, the reuse of areas degraded by quarrying for tourism purposes can be achieved by transforming the areas into theme parks, museums, spaces for environmental education, or even using the caves as a tourist attraction.

1.4.2. Inadequate disposal of solid waste

Improvements in socio-economic conditions combined with population growth have led to changes in human habits, such as superfluous consumption, and as a consequence there has been an increase in the generation of waste from human activities, since one of the stages of the production chain is based on the use of natural resources to generate a new product (TADA et al., 2003; FERREIRA et al., 2014).

Currently, there is a paradox with regard to concern about the depletion of natural resources and the preservation of the environment, given the media's great encouragement of increasingly accentuated consumerist habits, resulting in the consumption of the superfluous (FERNANDES, 2009).

Boscov (2008) comments that waste can be defined as any material discarded after or during the execution of industrial, commercial or domestic activities, for which appropriate disposal is required. Leão (1995) defines waste as a by-product of the production system that has a usable potential, although this potential is not utilised. On the other hand, the author continues, rubbish has no use and simply needs to be disposed of in a non-toxic and non-polluting way.

According to NBR No. 10.004, from the Brazilian Association of Technical Standards (ABNT). ABNT (2004),

> Solid waste is solid and semi-solid waste resulting from industrial, domestic,

hospital, commercial, agricultural, service and sweeping activities. This definition includes sludge from water treatment systems, that generated in pollution control equipment and installations, as well as certain liquids whose particularities make it unfeasible to discharge them into the public sewage system or bodies of water, or require solutions that are technically and economically unfeasible in view of the best available technology.

According to Art. 13 of Law 12.305/2010, municipal solid waste is classified according to its origin, and includes household waste, waste from sweeping, urban cleaning, among others. Its disposal is one of the biggest contemporary problems, the solution to which is centred on the economic, social, cultural and environmental segments, the latter being the main aspect to be considered (BRASIL, 2010). The solid waste production chain generates impacts on various scales, the effect of which must be assessed in order to minimise possible problems linked to their inadequate disposal (BOSCOV, 2008).

According to Padilla (2007), solid waste is classified according to its physical-chemical or infectious characteristics and emphasises the importance of knowing the origin of the waste or the processes that gave rise to it, in order to quantify the substances present in it and assess the possible risk to health or the environment.

1.4.2.1. Qualitative characterisation of waste

Ferraz (2008) comments that solid waste can be categorised according to its nature or origin:

I. Household and Commercial: includes waste collected from households, commercial establishments and others.

II. Sweeping: refers to waste from street sweeping services, public places and street markets, weeding, mowing and removal of waste not collected by the regular system.

III. Health: waste from hospitals, pharmacies, clinics, dental practices, laboratories, airports, bus terminals and expired medicines.

IV. Fairs and Markets: waste from street cleaning at municipal fairs and markets.

V. Rubble: this includes class II B waste, such as: earth, rubble from public and private land, excavations, demolitions, construction waste and material removed during the desilting of water bodies.

VI. Industrial: generated by industries, requiring the services of private individuals to collect their waste.

VII. Special and Hazardous: this includes waste from manhole cleaning, tree pruning, dead animal carcasses, commercial premises, households, abandoned vehicles, furniture in general, batteries, light bulbs, paints, among others.

With regard to the waste found in the study area, it is assumed that most of it was generated by households. Much of the waste was burnt, which made it difficult to identify, but the remaining waste had apparently been there for some time, such as packaging, plastic bags and pet bottles.

1.4.2.2. Main diseases related to inadequate waste disposal

During the study, few homes were observed in the places visited, but there was a predominance of improper waste disposal, which can have a negative impact on the health of the inhabitants themselves. With regard to the inadequate disposal of solid waste, Fernandes (2009) comments that one of the main consequences for human health is the proliferation of rodents and insects, the spread of which can cause illnesses when they come into contact with people. In addition, the rainy season can contribute to the development of mosquito larvae and vectors, resulting in diseases such as dengue.

The disposal of waste on the ground can lead to its contamination and consequently groundwater and surface water, due to the infiltration of leachate which, according to ABNT (1984), is a black liquid generated by the decomposition of organic matter that is acidic, foul-smelling and has a high polluting potential.

Other common problems linked to inadequate waste disposal include the proliferation of endemic diseases, environmental and visual pollution, and soil and groundwater contamination (FERNANDES, 2009).

Throughout the study, little was seen about the disposal of organic waste. This fact is associated with the small number of households located at the study site. In Brazil, the main legal references currently in force for the recycling of organic waste are as follows: Law No. 6894, of 16 December 1980, aimed at inspecting and supervising the production and trade of fertilisers, biofertilisers, among others; Decree No. 4.954, of 14 January 2004, which approves the regulation of the aforementioned law; CONAMA Resolution No. 375, of 29 August 2006, which deals with the criteria and procedures for the agricultural use, above all, of sewage sludge from Sewage Treatment Plants (STPs) and makes other provisions; as well as some normative instructions.

This waste can contain high levels of substances that are potentially harmful to living beings and the environment, which in turn can accumulate in living tissues and reflect levels that are dangerous to human health, such as mercury (Hg), which according to Py-Daniel (2013), is a heavy metal and can bioaccumulate in the food chain, which can have adverse effects on human health, especially on the development of foetuses and young children.

1.4.2.3. Permanent Preservation Area (APP)

Around the road there is a Permanent Preservation Area (APP) located along the body of water, which is approximately 10 metres wide and, according to Law No. 12,651 of 25 May 2012, which provides for the protection of vegetation, states in its Chapter II, Section I, Art. 4, **that the marginal strips of any natural perennial** and intermittent **watercourse**, excluding ephemeral ones, from the edge of the channel of the regular bed, in **minimum** width **according to the extension for the watercourses relating to their width.** The river is less than 10 metres wide and the vegetation is preserved, however, in this study it was not possible to verify whether it is 30 metres long and thus relate it to the provisions of the legislation in question.

Article 3, contained in the Forest Code, considers permanent preservation areas to be forests or forms of natural vegetation designed to contain land erosion, form buffer strips along roads and railways, protect sites of exceptional beauty and/or scientific or historical value, shelter specimens of fauna or flora threatened with extinction and also ensure the well-being of society.

Forests mitigate the process of soil erosion caused by the direct impact of rain on the soil. In addition, trees absorb and retain a large part of the nutrients that are essential for their proper development, so they have the capacity to recover degraded soils and areas. Other ecosystem services provided by forests are: the ability to stabilise the climate; increased infiltration of water into the soil, thus ensuring the recharge of underground aquifers and consequent conservation of biodiversity (MASCHIO et al., 2014; REDIN et al., 2011).

Burning was observed in the localities (1°8'23.003"S, 47°37'40.184"W), mostly on one side, which is associated with the construction of houses and possibly the cultivation of some species for local consumption.

The practice of burning the land for cultivation purposes is a type of traditional system. These systems release nutrients into the soil, especially calcium and magnesium, and these elements are closely linked to the pH of the environment, so the soil tends to become more basic, favouring the development of some species (REDIN et al., 2011). However, there is an undesirable effect on the soil's physical properties, such as porosity, where the volume of macropores is reduced and infiltration is noticeably altered, as a result of which density can increase, making it more difficult for roots to penetrate. In addition, unprotected soil is more susceptible to leaching and percolation of nutrients (STONE and SILVA, 2001; PANDOLFO et al., 2004).

1.4.2.4. Effluent into watercourses

Another frequent consequence of population growth is the pollution of water bodies as a result of improper dumping of human waste from various human activities (DOMINGOS and OLIVEIRA, 2008).

Pollution refers only to alterations in the chemical, physical and biological characteristics of the environment, while contamination, according to Federal Law 6938/81, contained in the National Environmental Policy (PNMA), as well as being related to alterations in its physical, chemical and biological properties, can result in damage to the health, safety and well-being of populations, cause harm to fauna and flora, or even jeopardise its use for economic or social purposes: urban waste, toxic substances and pesticides (FALAT et al., 2013).

For a coherent result regarding the discharge of effluents in the area, it is necessary to carry out analyses of certain parameters established in CONAMA Resolution 430/2011, such as biochemical oxygen demand (BOD) and hydrogen potential (pH), which can reflect the quality of the effluent discharged and consequently that of the water body (VON SPERLING, 2005). The aforementioned resolution "lays down conditions and standards for discharging effluents, complements and amends Resolution 357 of 17 March 2005 of the National Environment Council (CONAMA)". Chapter II, Article 16, establishes reference values for effluent discharge conditions. The pH must be between 5 and 9 and the BOD5 must have a minimum removal of 60%, and this limit can only be reduced if there is a self-depuration study of the water body that proves compliance with the goals of the framework of the receiving body.

1.5. Biodiversity

Analysing the biodiversity of a given location is important from a number of points of view, including economic, environmental, ethical and aesthetic maintenance. In general, biodiversity is understood as the multiplicity of species, whether plant or animal (MMA, 2002).

The biodiversity practical was carried out at UFRA's Igarapé-Açu School Farm. It consisted of setting up traps for birds and flying mammals, visualising and understanding the workings of a trap for capturing reptiles and amphibians, and searching for arthropods using a net to assemble an entomological box.

To capture birds and flying mammals, vertical nets were used in the late afternoon to catch the animals while they sought shelter. The nets were located in the middle of the trees, forming pockets to facilitate capture, as shown in figure 16. The aim of this practice is to monitor the species of birds and flying mammals, as well as their conditions, in order to make inferences about the conditions of the environment.

Figura 16:   (a/b) Bird capture by vertical nets.
Source: Authors, 2017.

The bird seized in the vertical net was a small thrush. After taking measurements, it was found to have a total length of 22.5 cm, a wing of 11.5 cm, a tarsus of 42.21 mm and a beak of 24.49 mm, weighing 69 g (Annex 3). Birds play an important role in ecosystems, including their ability to disperse seeds, helping to break their dormancy and genetic variability. The supply of resources in the environment will determine the presence of each species in it (CIAMBELLI, 2008).

Therefore, certain factors for each species will determine whether or not it remains in that environment, such as abundance, size, ecological function, among others. These characteristics will determine the resistance of each species to changes in the environment, with the more resistant species being able to remain, for example, in urban areas, depending on the resources found there. In this way, the large presence of the thrush in the environment, in relation to other birds, may indicate that the environment in which it is inserted is anthropised, which is what was observed on the Fazenda Escola. This is because this bird is more resistant to changes in the environment than other bird species (ANTAS and ALMEIDA, 2002; FILHO and SILVEIRA, 2012).

The same trap used to capture the bird was used to catch flying mammals. Bats make up 25 per cent of mammals and are very biodiverse. In terms of diet, they can be frugivores, insectivores, carnivores, omnivores and even haematophages. Most of them are insectivores, which is why bats play an important role in the biological control of pests that damage agricultural crops and insects that transmit diseases. These animals also disperse seeds, especially those that are fruit-eaters, favouring the recovery of degraded areas. Bats are excellent bioindicators and their characteristics and behaviour indicate when the environment is undergoing changes, as they are very sensitive to changes in the environment (PINA, 2011; TORRES, 2016).

The Artibeus lituratus species was captured at the Fazenda Escola at night (Figure 17). The general anatomy of some bat species, including the one seized, can be seen in Appendix 3. According to Brusco and Tozato (2009), this species is quite common in tropical regions, a fact explained by their diet, since they are frugivores. They are very important in environments affected by deforestation, for example, as they are able to disperse seeds over considerable distances, facilitating the regeneration of that environment.

Figura 17:   Artibeus lituratus species captured at the Igarapé-Açu School Farm.

Source: Authors, 2017.

Brusco and Tozato (2009) also state that dispersal can also occur in two ways: by epizoochory, where the animal's fur carries fruit or by the plant's own mechanisms for this to happen, such as sticky substances and hooks, or by endozoochory, which is done by ingestion and dispersal through faeces during flight. Novaes and Nobre (2009) comment that the diet of fruit bat species can vary according to the environment in which they are found. In conserved areas, for example, most of the fruit these animals consume is part of their diet, while in urban areas, they consume a lot of fruit from introduced species. It is therefore clear that they adapt to their surroundings.

During the field period, representatives of some orders of the phylum Arthropoda were also collected (Appendix 3), which were then organised in order to produce an entomological box with some of them (Figure 18), while others were kept in ethanol. These were captured using a hook and sinker in the vicinity of the Fazenda Escola during the afternoon and early evening.

Among the representatives captured was a spider, found at 3.03pm in an area of secondary vegetation, clearly anthropised, as it is interrupted by a small road (latitude 020998 and longitude 9875148, UTM 23S), with a hot and humid climate, where it showed aggressive behaviour when it was captured using a plastic container with wet cotton wool soaked in ethanol, according to Annex 3.

Figure 18: Assembling the entomological box.

Source: Authors, 2017.

Because they have a wide variety of species and are sensitive to environmental conditions and the organisation and layout of their habitat, spiders are also good bioindicators, signalling the conservation status of forest fragments and acting to control insects (ALVES et al., 2005). For example, it is possible to get an idea of the quality of leaf litter according to itinerant spiders on the ground. On the other hand, other species are influenced by the type of vegetation where they are found, and this factor is seen through the diversity of web construction (NOGUEIRA; PINTO-DA-ROCHA; BRESCOVIT, 2006).

1.6. Amazonian Agroecosystems

1.6.1. The country's agro-industrial scenario

The concept of agro-industrial systems (SAG) has vast applications, ranging from the design of public policies to the architecture of organisations and the development of corporate tactics. The Brazilian agro-industrial complex was strengthened at the end of the 1960s as a result of Brazilian industry's ability to produce goods and products for use in agriculture and livestock farming. The use of these goods led the agricultural-based sector to transformations that essentially incorporated management techniques, above all through the use of imported agricultural machinery (STAL et al., 2010; ZYLBERSZTAJN, 2000).

In Brazil, the palm oil agro-industrial sector has the world's greatest potential for producing palm oil, due to the almost 75 million hectares of land suitable for palm oil production.

The states of Pará, Bahia and Amapá stand out as the main producers of oil palm in the country (EMBRAPA, 2010).

The Agroecological-Economic Zoning of Oil Palm (ZAE-Dendê) established the Amazon region as a priority area for the expansion of oil palm plantations, with the Northeast of Pará being the area with the most suitable soil and climate conditions for this crop (EMBRAPA, 2010). According to Silva and Alves (2017), in terms of production value, oil palm is the third most important agricultural activity in northeast Pará, totalling R$274 million in 2014. These results for oil palm *(Elaeis gumeensis)* cultivation are growing due to specific public policies such as the National Programme for the Production and Use of Biodiesel (PNPB), launched in 2004 by the Federal Government, which is geared towards the production of agrofuel.

As noted above, the palm oil agro-industry has seen significant growth in recent years compared to other agro-industries, and this rise has been mainly due to government tax incentives. Furthermore, by analysing the information provided by the Palmasa company, it is possible to make a comparison between the productivity of palm oil and soya oil, which are one of the main vegetable oils in Brazil. It can be seen that palm oil is currently the most produced in the country, reaching around 3000 to 5000 kg of oil per hectare, while soya produces 400 to 600 kg of oil per hectare.

1.4.3. Versatility of the palm

Oil palm has a wide range of uses and can be applied in the food, cosmetics, pharmaceutical, oleochemical and biofuel industries. The wide variety of fractions obtained from the oil increases its use in various foods, such as margarine, ice-cream masses, chocolate flavourings, fats for frying, baking, biscuits, etc. In addition, the fractions obtained are also important for the oleochemical industry. Palm oil is important for the cosmetics industry because it contains many fatty acids, such as palmitic acid, stearic acid, oleic acid (omega 9) and linoleic acid (omega 6), as well as being a source of tocopherol and tocotrienol (vitamin E), which act as **antioxidants and are also rich in beta-carotene (D'AGOSTINI and GOIELLI, 2002).**

Palm oil is among the most qualified for biodiesel production, due to its composition, high productivity, low cost, production distributed throughout the year, regular and growing supply, as well as being grown in different areas, so there is no competition with other food crops.

3.6.3 Palmasa's history

Agroindustústria Palmasa SA, where the technical visit took place, is located in the municipality of Igarapé-Açu in the state of Pará. The company, initially called Agroindustrial Palmasa Ltda., was founded in 1986 by the municipality's Japanese colony and became a Sociedade Anônima (SA) in 1988, two years after it was founded. According to the information provided in the lecture given by the company's chemical engineer, the first plants began operating in July 1991, with funds from Sudam and Banco da Amazônia S.A., with 340 hectares of planted area and an initial processing capacity of 6 tonnes of bunches/hour, which has now been increased to 2,500 hectares planted and 20 tonnes of bunches/hour.

Palmasa is the only major agro-industry in the aforementioned municipality, whose local economy revolves around the Japanese colony, where palm oil, black pepper, passion fruit and livestock are grown (figure 19). The company's activity absorbs local labour, both in the urban population (Palmasa) and in the rural area in palm plantations and other areas. According to estimates by the agro-industry, 400 direct jobs are currently being generated (mill and field), indirectly generating income for around 2,000 people.

Figure 19: Panoramic view of Agroindústria Palmasa SA.

Source: Authors, 2017.

3.6.4   Methods used (mowing and fertilising)

In the process of planting oil palm, the company uses the mowing and fertilising method. Mowing the vegetation between the rows to simplify the establishment and development of livestock is carried out periodically in the first year. This practice in the cultivation of interspecific hybrid oil palm trees also aims to help with the operations that will be carried out in the plantation, making it possible to move around smoothly. These operations are

harvesting and transporting the bunches, fertilising, plant health control, lowering and mowing, etc (EMBRAPA, 2017).

The oil palm agro-industry produces a large number of by-products, including empty bunches (stalks), mesocarp fibre, palm kernel cake and effluents. Field supervisor Adriano reported during the visit that the by-products produced have high potential as organic fertiliser, due to their nutrient content. These by-products can be decomposed or taken directly to the plantation. In young plantations, care should be taken not to apply the undecomposed organic fertiliser close to the plant. In the case of production plantations, it should not be applied to the crown of the plant, as it hinders harvesting. It can be applied in heaps or distributed between plants in the planting line.

1.6.5.   Study area: Marajoara Farm I

PALMASA's area inspectors, Adriano and Alessandro, presented Fazenda Marajoara I, whose area, according to them, is equivalent to 375 ha, triangular in shape, with 143 plants per hectare, the spacing adopted being 7.8x9 and each plot corresponding to 28 plants. According to them, 40 per cent of the company's production goes to PALMASA and the remaining 60 per cent to third parties (partners). There are blocks of 20, 25 and 30 ha and within these blocks there are spaces for cars, and between the blocks, lanes for larger transports.

34

The area in question used to be pastureland and in 2004 oil palm planting began, with the aim, among other things, of providing an alternative way of recovering the area. The first oil palm is harvested after 4 years, but some producers use 3 years, depending on market demand and growing conditions (EMBRAPA, 2014).

During the visit, the area inspectors who helped with the technical visit commented that the main tools used to start the process of obtaining the oil palm bunches are: the sickle, the axe and the sack (with a steel tip and dimensions of approximately 1m to 1.5m), shown respectively in figure 20. Harvesting takes place every fortnight.

Figure 20: Basic tools for obtaining the product. From left to right you can see the sickle, the axe and the mattock respectively.

Source: Authors, 2017.

It was also explained that the ideal period for harvesting the fruit is a cycle of 10 to 12 days, when they are ripe. Increasing the cycle causes the fruit to ripen, which results in a loss of production. The time to remove the ripe bunch is when 3 to 4 fruits are detached from it; if this doesn't happen, the fruit is still green.

The ripe bunches go through quality control, which in turn affects oil production. For productivity purposes, Palmasa uses the following percentages: 70 per cent ripe bunch, 2 per cent green bunch and 28 per cent spent bunch. The empty bunch retains a lot of water and is generally used as organic fertiliser. When used on older plantations, this organic fertiliser is spread over the whole area mechanically, as all the roots are more spread out. In younger plantations, it is placed in specific places. It should be noted that employees prune once a year in older plantations, which is not the case in younger crops.

1.6.6. Productivity obtained by the agro-industry

In terms of productivity, this agro-industry obtains a total of 20 to 22 bunches per year. Oil palm plantations can generally be sustained for up to 25 years, and when they are older, it can make harvesting/production more difficult, as well as maximising the risks to workers, as the tools required will be larger due to the size of the trees, which can result in accidents at work.

1.6.7. Recurring problems in oil palm cultivation

The most common pests in this type of crop are: AF (fatal yellowing), which are commonly found in Brazil as a result of pioneering experiments with exotic African plants, the problem of which is being circumvented with the implementation of hybrids of national plants together with African ones, however it is not common in the region in question (EMBRAPA, 2014).

Another common pest is the nematode (Bursaphelenchus cocophilus) that causes the red ring on oil palm. The measure adopted to contain the adverse effects of this pest consists of using traps at the edges of the plantation, with pherormone and sugar cane, to attract the insect. Another recurring problem, especially during rainy periods, is lightning, which destroys the trees. It should be noted that oil palm cultivation has few (if any) problems with lizards, snakes and rodents.

For many years, entomophilous pollination, carried out by insects, was forgotten, as the wind was considered the only pollinating agent. This was certainly because the inflorescences had characteristics peculiar to wind-pollinated species, such as abundant pollen production (TUNER and GILBANKS, 1974; MOURA, 2008). On the Marajoara farm, a large number of beetles were found, which are responsible for the natural pollination of the oil palm.

For some years now, this agro-industry has not used herbicides which, according to Guimarães (1987), have their behaviour conditioned by various physical, chemical and biological factors and processes that cause the product to degrade and move. In the case of soils, the author continues, herbicides are transported in the soil mainly through pore spaces, diffusion in soil water, downward flow of soil water and upward movement of soil water. The company also doesn't use chemical products, as these can eliminate beneficial insects from the plantation, damaging the natural pollination done mainly by beetles.

According to Embrapa (1987), pruning is relevant in economic terms, in order to obtain a more profitable harvest, and confers benefits such as better natural fertilisation of clean, well-aerated inflorescence crowns and reduced exposure to labour risks. However, oil palm cultivation, like any other crop, also suffers from stress, which can result both from excessive pruning, especially in the less rainy periods, and from transport movements, causing soil compaction. As a result of this stress, the plant may develop more male inflorescences and fewer female ones, aborting the bunches and resulting in lower production. The climate also has a major impact on oil palm plantations, and may be responsible for the bunches aborting.

It was also explained that the pickers receive R$0.25 for each ripe bunch plus their salary. A green bunch is equivalent to less than 10 ripe bunches harvested (penalised). Some receive up to R$80.00 per harvested bunch.

It was mentioned that other companies carry out artificial pollination (removing the male inflorescence) of the oil palm, which can be beneficial to the crop, depending on the circumstances. Another relevant aspect that was expressed was the oil palm plantation, which currently accounts for 80 per cent of the food sector, where there is the greatest domestic demand. Palm is made up of oil and fat, and fractionation is at 45°C. Palm degradation is slower due to the presence of antioxidants.

1.6.8. Processing methods and by-products obtained

The processing of oil palm production must begin immediately after harvest and can be summarised in the following steps:

1. Fruit bunch: obtained from the cultivation of palm oil.
2. Sterilisation: consists of cooking the fruit at a temperature of 150 °C, varying in three bars, using steam only, with the aim of inactivating enzymes that cause acidity, as well as facilitating the detachment of the fruit from the bunches, thus causing the oil-containing cells to rupture. This process paralyses the microbiological action.
3. Threshing: separating the empty bunch from the fruit. It is worth noting that the empty bunch is returned to the field for organic fertiliser (composting) through its decomposition.
4. Digestion: maceration to release the oil in the pulp, subjected to a temperature of 95 °C.
5. Pressing: Raw (in natura), separating the fibre, which is used to generate heat for the factory, or nuts, to generate another oil, palm kernel oil.
6. CPO clarification: the industry's goal is to have up to 3 per cent acidity, but it currently has around 2.5 per cent. It can generate two alternatives: CPO storage and CPO cake.
   6.1 CPO storage: conditioning the by-product.
   6.2 CPO Pie: consists of the husks and fibres of the mesocarp, which have energy potential. It is also used to feed livestock.

Palm kernel oil, also known as palm kernel oil or coconut oil, is extracted from the kernel found inside the palm tree. This oil contains predominantly two acids (lauric and myristic), which justifies its use in the manufacture of soaps, detergents, ointments, mayonnaise, as well as its inclusion in chocolate production as a substitute for cocoa butter (SANTOS et al., 2014).

According to the chemical engineer at the PALMASA agro-industrial plant, the palm kernel processing process takes place as follows: a) nuts; b) breakers; c) separation (shells and energy source); d) drying of kernels; e) milling; f) cooking; g) pressing (PKO cake, used to feed pigs); h) PKO filtration (PKO storage).

Oil palm production generates a number of by-products, as illustrated in figure 21 below, with emphasis on CPO cake, made from the palm, and PKO cake, made from palm kernel, both of which are used to feed cattle.

7. Figure 21: By-products generated by palm processing. a) CPO cake. b) KPO cake.

Source: Authors, 2017.

The Palmasa agro-industrial company has a quality control system, which covers: controlling and monitoring the quality of the raw material; controlling and monitoring the quality of the inputs; controlling and monitoring the quality of the process; monitoring industrial efficiency; controlling and monitoring the quality of the end product and also providing customer support.

All the fruit is left in the courtyard (the start of the process), the fruit goes to the ramp, passes through conveyor belts and finally goes to the sterilisation room for approximately 90 minutes. Then they go to the tractors and on to the threshers.

In this company, the most common accidents are predominantly related to incorrect handling of basic harvesting tools such as axes, scythes and weeding tools.) A very widespread piece of information that is passed on to employees is that they should be careful when the leaves fall, as they contain thorns.

Pruned leaves and stems are left on the path (figure 22) in order to promote organic fertiliser through their decomposition, but taking care with their thorns so that no worker runs the risk of an accident.

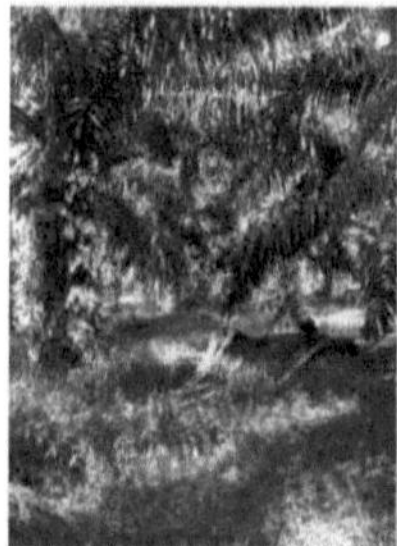

8. Figure 22: Pruned leaves and straw left on the path for organic fertiliser.

Source: Authors, 2017.

3.6.9 Effluent treatment

As explained during the technical visit, what is not treated is used for fertigation and what comes out of the reactors can be sold to make soap, with a COD load of 1000 ppm. The method used to treat effluents is the polishing lagoon and, according to Rodrigues et al. (2009), this type of treatment combines lower oxygen demand with higher oxygen production, thus predominating an aerobic environment, reflecting the prevalence of photosynthesis over bacterial oxidation.

CHAPTER 4

FINAL CONSIDERATIONS

The municipality of Igarapé-Açu has been developing positively over the last few years, and today it has gained economic prominence mainly due to its oil palm plantations and the processing of this species, which generates various commercial by-products. The growth of agro-ecosystems in the region is driven by the strong demand from the national market that the versatile use of oil palm has provided.

In relation to the geoprocessing analysis carried out with the aid of remote sensing in the areas visited, the satellite images showed worrying levels of degradation, with the presence of areas in clay, deforested and polluted environmentally and visually by the presence of various types of waste. Monitoring through the use of geotechnologies allows better visualisation of the consequences generated by the use and occupation of areas and facilitates the mapping of critical points and points of risk for the environment and the surrounding community.

The vegetation identified in the area is characterised as secondary forest of the capoeira type, which is the result of anthropogenic disturbances to the primary vegetation, which occurs through a process of natural regeneration. The existing biodiversity in the study area of this report is effective in providing good indications of the degree of anthropisation in the region, and

can be used through more in-depth studies to point out relevant impacts on the ecosystem balance.

For the soils observed in the region, which are heavily weathered, despite the obvious degradation, natural regeneration was observed through secondary vegetation, helping in the process of recovering the site's natural properties. However, there is a need for intervening actions to speed up this recovery process, such as the adoption of PRADAS - Projects for the Recovery of Degraded Areas, through the government agencies present in the region.

The limnological analyses obtained at the site made it possible to better understand the physical aspects and hydrographic dynamics of the municipality and to realise that the public bathing area studied, which is very popular in this region, does not yet show significant changes that compromise the quality of use of these waters.In view of the above, this work enabled students on the Environmental Engineering course at UFRA to face the challenges of carrying out an extensive survey of a region in the north-east of Pará and to assess its environmental reality by applying skills learnt in the classroom in the field.

CHAPTER 5

REFERENCES

ALBUQUERQUE, M. F. et al. Precipitation in the Mesoregions of the State of Pará: Climatology, Variability and Trends in Recent Decades (1978-2008). **Revista Brasileira de Climatologia,** Curitiba, v. 6, n. 6, p.151-168, Not a valid month! 2010. ISSN: 1980-055X. Available at: <http://revistas.ufpr.br/revistaabclima/article/view/25606>. Accessed on: 05 October 2017.

ALHO, C. J. R. Importance of biodiversity for human health: an ecological perspective. **Estud. av.,** São Paulo, v. 26, n. 74, 2012. Available at: http://www.scielo.br/scielo.php?script=sci_arttext&pid=S010340142012000100011&lng=pt &nrm=iso. Accessed on: 28 September 2017.

ALMEIDA, C. A. et al. Estimation of secondary vegetation area in the Brazilian Legal Amazon. **Acta Amazônia.** v. 40. p. 289-302. 2010. Available at: < http://www.scielo.br/pdf/aa/v40n2/v40n2a07>. Accessed: 4 October 2017.

ALMEIDA, C. A. et al. Methodology for mapping secondary vegetation in the legal Amazon. **Ministry of Science and Technology: INPE.** Available at: <:https:// www.academia.edu>. Accessed: 2 October 2017.

ALVES, A. O. et al. Study of spider communities (Arachnida: araneae) in an Atlantic forest environment in the Pituaçu Metropolitan Park - PMP, Salvador, Bahia. **Biota Neotropica,** v. 5, p. 1-8. Salvador, 2005. Available at: < http://www.scielo.br/pdf/bn/v5n1a/v5n1aa07.pdf >. Accessed on: 3 October 2017.

ANTAS, P. T. Z.; ALMEIDA, A. C. Birds as bioindicators of environmental quality. **Aracruz CeluloseS/A.** 2002. Available at :< https://www.researchgate.net/profile/Paulo_Antas/>. Accessed on: 1 October 2017.

BRAZILIAN ASSOCIATION OF TECHNICAL STANDARDS. NBR 10004. Rio de Janeiro, 2004.

BAENA, A. R. C.; FALESI, I. C.; DUTRA, S. Physical-chemical characteristics of the soil in different agroecosystems in the Bragantina region of north-eastern Pará. Belém: **Embrapa CPATU,** 1998. Available at : < http://ainfo.cnptia.embrapa.br/digital/bitstream/item/48885/1/Boletim-Pesquisa-185- CPATU.pdf >. Accessed on: 29 September 2017.

BARBIERI, E. **Biodiversity: The Variety of Life on Planet Earth.** São Paulo. 2010.

BOSCOV, Maria Eugenia Gimenez. **Environmental Geotechnics.** Text workshop. São Paulo, 2008.

BRAZIL. Law n. 12. 305, of 2 August 2010. Establishes the National Solid Waste Policy. Amends Law No. 9.605, of 12 February 1998; and makes other provisions. National Solid Waste Policy.

BRAZIL. Law no. 12.651, of 25 May 2012. Provides for the protection of native vegetation; amends Laws Nos. 6.938, of 31 August 1981, 9.393, of 19 December 1996, and 11.428, of 22 December 2006; repeals Laws Nos. 4.771, of 15 September 1965, and 7.754, of 14 April 1989, and Provisional Measure No. 2.166-67, of 24 August 2001; and makes other provisions. Office of the Chief of Staff for Legal Affairs.

BRUSCO, A. R.; TOZATO, H. C. Frugivory in the diet of Artibeus lituratus Olfers, 1818 (Chiroptera, Phyllostomidae) in Parque do Ingá, Maringá / PR. **Revista FAP Ciência.** v.3, n. 2, p. 19 - 29, Apucarana, 2009. Available at: < http://www.cesuap.edu.br/fap- ciencia/edicao_2009/002.pdf >. Accessed on: 5 October 2017.

CIAMBELLI, C. P. Survey of birds and their contribution to the recovery of the Botucatu State Forest - Botucatu/SP. Monograph (Bachelor's Degree in Biological Sciences) - São Paulo State University, 2008.

Available at: <https://repositorio.unesp.br/bitstream/handle/11449/118677/ciambelli_cp_tcc_bot.pdf?seque nce=1>. Accessed on: 2 October 2017.

CONAMA - National Environment Council. **Resolution No. 357 of 17 March 2005,** available at: <http://www.mma.gov.br/port/conama/res/ res05/res35705.pdf>. Accessed on: 05 October 2017.

CORTEZ, C. M. B.; TAGLIARINI, E. M.; TANCREDI, A. C. F. N. S. Utilisation of Mineral Waters from the Barreiras Group Aquifers in the Region of Belém (Pa). 2000. Available at: <https://aguassubterraneas.abas.org/asubterraneas/article/viewFile/23751/15818>. Accessed on: 26 September 2017.

COSTA, J.B.S.; BORGES, M.S.; BEMERGUY, R.L.; IGREJA, H.L.S.; PINHEIRO, R.V.L. Aspects of Cenozoic tectonics in the Salgado region, NE coast of the State of Pará. In: **SIMPÓSIO DE GEOLOGIA DA AMAZÔNIA,** 3.Belém. Brazilian Geological Society, Belém. 1991. P. 156-165.

COSTA, P. R. C.; LIMA, E. A. Brazilian Symposium on Chemistry and Pharmacology of Natural Products. Rio de Janeiro, 1989.

DAINESE, R. C. Remote Sensing and Geoprocessing Applied to the Temporal Study of Land Use and the Comparison between Non-Supervised Classification and Visual Analysis. 2001. 210 f. Dissertation (Master's Degree) - Agronomy Course, Department of Rural Engineering, Faculty of Agronomic Sciences, Faculty of Agronomic Sciences, Unesp, Botucatu - Sp, 2001. Available at: <http://200.145.6.238/bitstream/handle/11449/90651/dainese_rc_me_botfca.pdf?sequence=1 &isAllowed=y>. Accessed on: 06 October 2017.

DOMINGOS, E. Q.; DE OLIVEIRA, E. M. S. Evaluation of domestic effluent treatment systems in the municipality of conceição de macabu-rj. 2008.

DRIESSEN, P.M. & DUDAL, R. Lecture notes on the geography, formation, properties and use of the major soils of the world. **Wageningen, Agricultural University,** p. 296, 1989.

Embrapa 2008. **Agrometeorological Bulletin** 2005: Igarapé Açu, Pa. Available at: <https://www.infoteca.cnptia.embrapa.br. Accessed: 27 September 2017.

EMBRAPA. EMPRESA BRASILEIRA DE PESQUISA AGROPECUÁRIA National Soil Research Centre. **Brazilian Soil Classification System.** Rio de Janeiro, Embrapa Solos. 2nd Ed. 2006. 306 p.

EMBRAPA. BRAZILIAN AGRICULTURAL RESEARCH COMPANY. National Soil Research Centre. **Brazilian Soil Classification System.** Brasilia, Embrapa. Produção de Informações; Rio de Janeiro, Embrapa Solos 1999. 412p.

EMBRAPA. Brazilian Agricultural Research Corporation. National Soil Research Centre. **Brazilian Soil Classification System.** Brasilia, Embrapa. Produção de Informações; Rio de Janeiro, Embrapa Solos 1999. 412p.

EMBRAPA. BRAZILIAN AGRICULTURAL RESEARCH COMPANY. **Manual for Training Community Groups in Participatory Water Quality Monitoring Methodologies.** Documents 135. ISSN 2179-8184. Fortaleza, CE. 1st ed. 2011. 48p.

ESTEVES, F. A. (Coord). **Fundamentals of liminology.** Rio de Janeiro: Interciencia, 3 ed. 2011. 826 p.

FALAT, S.C; DIFFONTE, S.C.R; WEBER, M.I. Disposal of solid urban waste. **Tuiuti University of Paraná.** 2013

FALESI, I. C.; GALEÃO, R. R. **Recovery of anthropised areas in the Northeast Mesoregion of Para through Agroforestry Systems.** In: IV Brazilian Congress on Agroforestry Systems: Agroforestry Systems, a trend in ecological agriculture in the tropics: sustaining life and sustaining life. Ilhéus - BA, 2004, 292p

FAPESPA - Fundação Amazônia de Amparo a Estudos e Pesquisas. **Municipal Statistics for Pará: Igarapé - Açu**. Directorate of Statistics and Information Technology and Management. Belém, 59f. n. 1, Jul/Dec 2016.

FERNANDES, D. N. The inadequate management of urban solid waste in the community of prado, neighbourhood of catolé, campina grande/PB. **Revista OKARA**: Geografia em debate, v.3, n.2, p. 223-347, 2009.

FERRAZ, José Lázaro. **Modelo para Avaliação da Gestão Municipal Integrada de Resíduos Sólidos Urbanos**. 2008, 241p, Thesis (Doctorate in Mechanical Engineering) - State University of Campinas, Campinas, 2008.

FERREIRA, E. M. et al. Final disposal of urban solid waste: diagnosis of management in the municipality of Santo Antônio de Goiás. **Revista Monografias Ambientais - REMOA** v.14, n.3, p.3401-3411, 2014.

FERREIRA, W.C; BOTELHO, S.A; DAVIDE, A.C; FARIA, J.M; FERREIRA, D.F. Natural regeneration as an indicator of the recovery of a degraded area downstream of the Camargos hydroelectric plant, MG. **Revista Árvore**, n. 34,p. 651-660, 2010

FILHO, J. C. M.; SILVEIRA, R. V. Composition and trophic structure of the bird community in an anthropised area in the west of the state of São Paulo. **Ornithological News.** [On-line]. n. 169, p. 33-40. São Paulo, 2012. Available at:< http://www.ao.com.br/download/AO169_33.pdf>. Accessed on: 2 October 2017.

FLORENZANO, T. G. Geomorfologia: conceitos e tecnologias actuais. 2ª ed. São Paulo - SP: Oficina de Textos, 2008.

GODOI, S; M, TAKAKI. Effect of temperature and the participation of phytochrome in the control of embauba seed germination. **Revista Brasileira de Sementes**, vol. 27, n° 2, p.87-90, 2005.

GOÉS, A.M.; ROSSETTI, L. F.; NOGUEIRA, A. C. R.; TOLEDO, P. M. **Preliminary depositional model of the Pirabas Formation in north-eastern Pará State**. Belém. 1990. Bulletin of the Emílio Goeldi Museum of Pará. Belém. 1990. 2: 3 - 15

IBGE - BRAZILIAN INSTITUTE OF GEOGRAPHY AND STATISTICS. **Panorama by municipality: 2017**. Rio de Janeiro: IBGE, 2017. v. 4.2.4. Available at: <https://cidades.ibge.gov.br/brasil/pa/igarape-acu/panorama>. Accessed on: 27 September 2017.

IBGE 2000;   2010 IBGE. **Demographic   Census-2010,**   available at: <https://www.ibge.gov.br/> Accessed on: 28 Sep. 2017.

IGREJA, H. L. S.; BORGES, M. S.; ALVES, R. J.; COSTA JÚNIOR, P. S; COSTA J. B. S. Neozonic studies on the islands of Outeiro and Mosqueiro - NE of the State of Pará. In: CONGRESS OF GEOLOGY, 33. 1990. Natal. Brazilian Geological Society. Natal 1990. Proceedings: 5: 2: 2110 -2113.

JÚNIOR, Antônio Carlos Ribeiro Araújo. Indicators of Environmental Quality in Lake Bologna, Utinga State Park, Belém-Pará. **Boletim Gaúcho de Geografia**, Porto Alegre, v. 42, n. 1,   p.276-299,   Jan,   2015. Available at: <http://www.seer.ufrgs.br/index.php/bgg/article/view/48373/32948>. Accessed on: 05 October 2017.

KASPER, D.; Sarah S. Bioaccumulation of mercury in piscivorous fish from the balbina reservoir, amazonas. 2013.

KATO, O. R. et al. Use of Agroforestry in the management of secondary forests. In: Agroforestry systems: scientific bases for sustainable development. 1. ed. Campos dos Goytacazes: Universidade Estadual do Norte Fluminense Darcy Ribeiro, 2006. P. 119-138.

KELLER, P. F. The artisan and the craft economy in contemporary society. **Revista de Ciências Sociais**, n. 41, p. 323-347, 2014.

KOPEZINSKI, Isaac. Mining X Environment: legal considerations, main environmental impacts and their modifying processes. Porto Alegre: UFRGS, 2000.

KUNZ, S. H. et al. Analysis of floristic similarity between forests of the Upper Xingu River, the Amazon Basin and the Central Plateau. **Revista Brasil.** v. 32. p. 725-736. 2009. Available at: < https://s3.amazonaws.com/academia.edu.documents>. Accessed: 9 October 2017.

LEANDRO, L. M. L.; SILVA, F. C. The Bragança railway and the colonisation of the Bragantina area in the state of Pará. **Novos Cadernos NAEA**, v. 15 n. 2, p. 143-174, dec, 2012.

LEÃO, Antônio Leandro. **Renewable Natural Resources**. São Paulo: CETESB, 1995, 93p.

LOPES, M. N. G.; SOUZA, E. B. de; FERREIRA, D. B. da S.. Regional Climatology of Precipitation in the State of Pará. **Revista Brasileira de Climatologia**, Curitiba, v. 12, n. 9, p.84- 102, jan. 2013. Quarterly. ISSN: 1980-055x (Print) 2237-8642 (Electronic). Available at: <http://revistas.ufpr.br/revistaabclima/article/view/31402/21508>. Accessed on: 05 October 2017.

MAGALHÃES, C. F. Diretrizes para o turismo sustentável em municípios. São Paulo: Roca, 2002.

MASCHIO, M. Soil contamination due to incorrect waste disposal

MCE - MINISTRY OF SCIENCE AND TECHNOLOGY. Researchers study the risk of oil palm expansion in Pará. Brasília: 2013. Available at: <http://saturno.museu-goeldi.br/inct/index.php?option=com_content&view=article&id=159>. Accessed on: 27 September 2017.

MESQUITA, R. A. et al. The importance of permanent preservation areas (APP'S).

MIRANDA, R. R. **Public Policies for Rural Development, Technological Innovations and Peasant Territoriality in Igarapé-Açu - Pa/Brazil.** III National Symposium on Agrarian Geography - II International Symposium on Agrarian Geography Jornada Ariovaldo Umbelino de Oliveira. Presidente Prudente, 2005.

MMA - MINISTRY OF THE ENVIRONMENT. Evaluation and identification of priority areas and actions for the conservation, sustainable use and sharing of the benefits of biodiversity in Brazilian biomes. p. 404. Brasília: MMA/SBF, 2002. Available at: <http://www.mma.gov.br/estruturas/chm/_arquivos/biodivbr.pdf>. Accessed on: 1 October 2017.

MMA - MINISTRY OF THE ENVIRONMENT. Brazilian Biodiversity: Assessment and identification of priority areas and actions for the conservation, sustainable use and sharing of the benefits of biodiversity in Brazilian biomes. Brasília: MMA/SBF, p. 12-13, 2002.

MOREIRA, H. L.; OLIVEIRA, V. A. Evolution and genesis of a Plintossolo Pétrico concrecionário êutrico argissólico in the municipality of Ouro Verde de Goiás. **Revista Brasileira de Ciência do Solo**, v. 32, n. 4, p. 1683-1690, 2008.

MOTA, R.V ; DE FRANÇA, L. F. Study of the characteristics of ucuuba (virola surinamensis) and inajá (maximiliana regia) with a view to biodiesel production. Revista cientifica da UFPA, n. 01, v. 06, 2007.

NIEMI, G.J. & MCDONALD, M.E. Application of ecological indicators. **Annual Review of Ecology, Evolution Systematics**, v.35, p. 89-111, 2004.

NOGUEIRA, A. A.; PINTO-DA-ROCHA, R.; BRESCOVIT, A. D. Community of orb-weaving spiders (Araneae, Arachnida) in the Morro Grande Forest Reserve, Cotia, São Paulo, Brazil. **Biota Neotrop.** v. 6, n. 2.

São Paulo, 2006. Available at: < http://www.scielo.br/pdf/bn/v6n2/v6n2a09.pdf >. Accessed on: 2 October 2017.

NOVAES, R. L. M.; NOBRE, C. C. Diet of Artibeus lituratus (Olfers, 1818) in an urban area in the city of Rio de Janeiro: frugivory and a new record of folivory. **Chiroptera Neotropical**. Rio de Janeiro, 2009. Available at :< https://www.researchgate.net/profile/Roberto_Leonan_Novaes/ >. Accessed on: 5 October 2017.

NUNES, A. N.; ALMEIDA, A. C.; COELHO, C. O. A. Impacts of land use and cover type on runoff

OLIVEIRA NETO, Manoel Batista de; SILVA, Maria Sonia Lopes da. **Latossolos Amarelos**. 2011. Prepared by EMBRAPA. Available at: <http://www.agencia.cnptia.embrapa.br/gestor/territorio_mata_sul_pernambucana/arvore/CO NT000gt7eon7j02wx7ha087apz21f7726p.html>. Accessed on: 03 October 2017.

OLIVEIRA, J.B. Pedologia aplicada. Jaboticabal, FUNEP, 2001. 414p.

OLIVEIRA, L. C. K.; CANELLAS A. V. B. **Importance of Reliable Hydrometeorological Data in Water Resources Management**. Proceedings of the International Symposium on Water Resources Management. Gramado, RS. 1998.

ONZONI, F. J.; ALMEIDA, E. Sper de. **Estimating the Kappa Parameter (K) of Discrete Multivariate Analysis in the Context of a GIS**. In: VIII Brazilian Symposium on Remote Sensing, 1996, Salvador. Proceedings. Salvador: INPE, 1996. p. 729 - 733. Available at: <http://marte.sid.inpe.br/col/sid.inpe.br/deise/1999/02.05.09.39/doc/T130.pdf>. Accessed on: 06 October 2017.

PACHÊCO, N. A.; BASTOS, T. X. Agrometeorological Bulletin 2006 - Igarapé-Açu. **Belém: Embrapa Amazônia Oriental,** 2007. Available at :< http://www.cpatu.embrapa.br/publicacoes_online >. Accessed on: 27 September 2017.

PANDOLFO, C. M. et al. Edaphic fauna in soil management systems and nutrient sources. In: Brazilian Meeting of Soil Fertility and Plant Nutrition, Lages, SC. **Proceedings**... Lages: SBCS/SBM, 2004.

PINA, S.M.S. Habitat use by bats in natural and agroforestry systems in the Brazilian savannah. Dissertation (Master's Degree in Ecology and Environmental Management) - University of Lisbon, 2011 : <http://repositorio.ul.pt/bitstream/10451/4651/1/ulfc090944_tm_sonia_pina.pdf>. Accessed on: 5 October 2017.

PORTELA, B. T. T. et al. Water balance of the municipality of Igarapé Açu. Available at:< https://www.cbtm.com>. Accessed: 5 October 2017.

PORTELA, B. T. T.; SOUZA, D. C.; SOUZA, P. J. O. P. **Water Balance of the Municipality of Igarapé-Açu - PA**. (Poster presentation). In: XIV Brazilian Congress of Meteorology, 2006, Florianópolis. Proceedings of the XIV Congress of Meteorology, 2006.

PY-DANIEL, S.S; KASPER, D; FORSBERG, B.R. Bioaccumulation of mercury in piscivorous fish from the balbina reservoir, amazonas. **II PIBIC/CNPq Scientific Initiation Congress**. Manaus, 2013.

REDIN, M. et al. Impacts of burning on chemical, physical and biological soil attributes. **Ciência Florestal, Santa Maria**, v. 21, n. 2, p. 381-392, 2011.

RIBEIRO, H. M. C.; PICANÇO, A. R. S.; CRUZ, L. D. F. **Environmental Analysis of Water Quality in Lake**

Bologna, Belém-PA, Brazil. **In:** Anais...50° Congresso Brasileiro de Química. 2010.

SANTOS, I.A. JÚNIOR, S.B. ALEGRE, J. **Fireless Agriculture in Eastern Amazonia: Ants as Agroecological Indicators. Biological**, São Paulo, v.69, supplement 2, p.53-56, 2007.

SECCO, D.; REINERT, D.J.; REICHERT, J.M. & DA ROS, C.O. Soya productivity and physical properties of a Latosol submitted to management systems and compaction. **R. Bras. Ci. Soil**, 28:797-804, 2004.

SILVA E. L. **Perspectivas De Desenvolvimento Municipal Em Igarapé-Açu, Pará, Brasil.** Dissertation (Master's in Natural Resource Management and Development - Federal University of Pará, Centre for the Environment - NUMA) Belém 2010.

SILVA, B. N. R. et al. **Soils of the Hydrographic Mesobasins of the São João and Cumaru Igarapés, Municipalities of Marapanim and Igarapé Açu.** In: XXXII Brazilian Congress of **Soil Science: "Soil and Bioenergy Production: Perspectives and Challenges".** Brazilian **Society of** Soil Science, 2012, Fortaleza. Proceedings of the XXXII Brazilian Congress of Soil Science.

SILVA, B. N. R.; SILVA, L.G. T.; RODRIGUES, T. E.; GERHARD, P. **Solos das Mesobacias Hidrográficas dos Igarapés São João e Cumaru, Municípios de Marapanim e Igarapé Açu.** Proceedings...XXXII Brazilian Congress of Soil Science. Brazilian Society of Soil Science. Fortaleza. 2009.

SILVA, Benedito Nelson Rodrigues da et al. **Soils of the Hydrographic Mesobasins of the São João and Cumaru Igarapés, Municipalities of Marapanim and Igarapé Açu.** 2009. XXXII **Brazilian Congress of Soil Science: "Soil** and Bioenergy Production: Perspectives and **Challenges". Brazilian Society of Soil Science. Fortaleza. Available at:** <ttps://www.alice.cnptia.embrapa.br/bitstream/doc/572594/1/2419.pdf>. Accessed on: 05 October 2017.

SILVA, I. M. et al. Recovering the Management and Conservation of Hydrographic Microbasins in Igarapé Açu (Pa): Considerations on the use of Agroforestry Systems. **SOBER.** Available at:< https://citeweb.inf>. Accessed: 6 October 2017.

SILVA, L. G. T. et al. **Soil Mapping in Two Hydrographic Mesobasins in Northeastern Pará.** Embrapa Amazônia Oriental (CPATU), 2013. 38p. Document (Embrapa Amazônia Oriental. Documentos, 394). Available at: <http://ainfo.cnptia.embrapa.br/digital/bitstream/item/97029/1/DOC-394.pdf>. Accessed on: 26 September 2017.

SILVA, M. L. **Palaeosols and Quaternary Environmental Studies: Theoretical Discussion and Possible Applications.** Brazilian Journal of Physical Geography, 2011.

SILVEIRA, C. S.; SOUZA, K. V. **Hydrological Relationships between Rainfall and Flow a Time Series (2007-2009) of a Mixed Use Drainage Basin - Teresópolis, RJ, BRAZIL.** São Paulo, UNESP, **Geosciences,** v. 31, n. 3, p. 395-410, 2012.

Solids in permanent preservation areas (APP). **International Symposium on Environmental Quality.** Porto Alegre, 2014.

SOUZA, C. R. G. et al. **Quaternary of Brazil.** Ribeirão Preto, SP: Holos Editora, 2005.

STONE, L. F.; SILVEIRA, P. M. Effects of tillage system and crop rotation on soil porosity and density. **Revista Brasileira de Ciência do Solo,** Viçosa, v. 25, n. 2, p. 395-401, 2001.

TADA, Agnes et al. **Urban solid waste:** sustainable landfill for small municipalities. Florianópolis, 2003.

TAMASAUSKAS, P. F. L. F.; TAMASAUSKAS, C. E. P. **Changes in Land Use and Cover and Surface Runoff in the Caripi-Pa River Basin: An Analysis Based on Geotechnologies.** GeoAmazonia Journal. ISSN: 2358-1778 (on line), Belém, v. 04, n. 08, p. 153 - 173, jul./dez. 2016.

TORRES, J.M. Biomonitoring of a large congregation of bats in the Catimbau National Park, Pernambuco. Dissertation (Master's in Animal Biology) - Federal University of Pernambuco, 2016. Available at: < http://repositorio.ufpe.br/bitstream/handle/123456789/17456/Disserta%C3%A7%C3%A3o% 20Corrigida%20Jaire%20Marinho%20Torres.pdf?sequence=1&isAllowed=y>. Accessed on: 4 October 2017.

TRINDADE, I. A.; SILVA, P. S. S.; SILVA, L. O. A. **Influence of Amazon Forest Deforestation in Igarapé Açu, PA.** IV Symposium on Research in Environmental Sciences in the Amazon. Belém, Pa. 2015.

VALENTE, Moacir Azevedo et al. **Soil Mapping of the Areas of Two Hydrographic Mesobasins in Northeastern Pará.** In: XXXII Brazilian Congress of Soil Science, 2011.   UberlândiaMG   .
                Disponívelem   :
<https://ainfo.cnptia.embrapa.br/digital/bitstream/item/59916/1/Ok-PA-CBCS33-Moacir-a-   Valente.pdf>. Accessed on: 26 September 2017.

VANZIN, M. M. **Evaluation of the sustainable use of water in agricultural production: impact of the insertion of agroforestry systems in family production units in north-eastern Pará.** 2014. Dissertation (Master's Degree in Family Farming and Sustainable Development) - Federal University of Pará, Belém, 2014.

VENTURIERI, G. C. **Conservation and Income Generation: Meliponiculture Among Family Farmers in Eastern Amazonia .** VII meeting on bees Ribeirão Preto, 2006.

VIEIRA, T. A.; ROSA, L.S.; VASCONCELOS, P. C. S.; SANTOS, M.M.; MODESTO, R. S. Adoption of Agroforestry Systems in Family Farming in Igarapé-Açu, Pará, Brazil **Rev. ciênc. agrár.** Belém, n. 47, p. 9-22, jan/jun. 2007.

VON SPERLING, M. Princípio do Tratamento Biológico de Águas Residuárias. Belo Horizonte, ed: 3, p. 452, UFMG, 2005.

WAICHMAN, A. V. **Water Quality.** Belém, Pa. 28 p.2002.

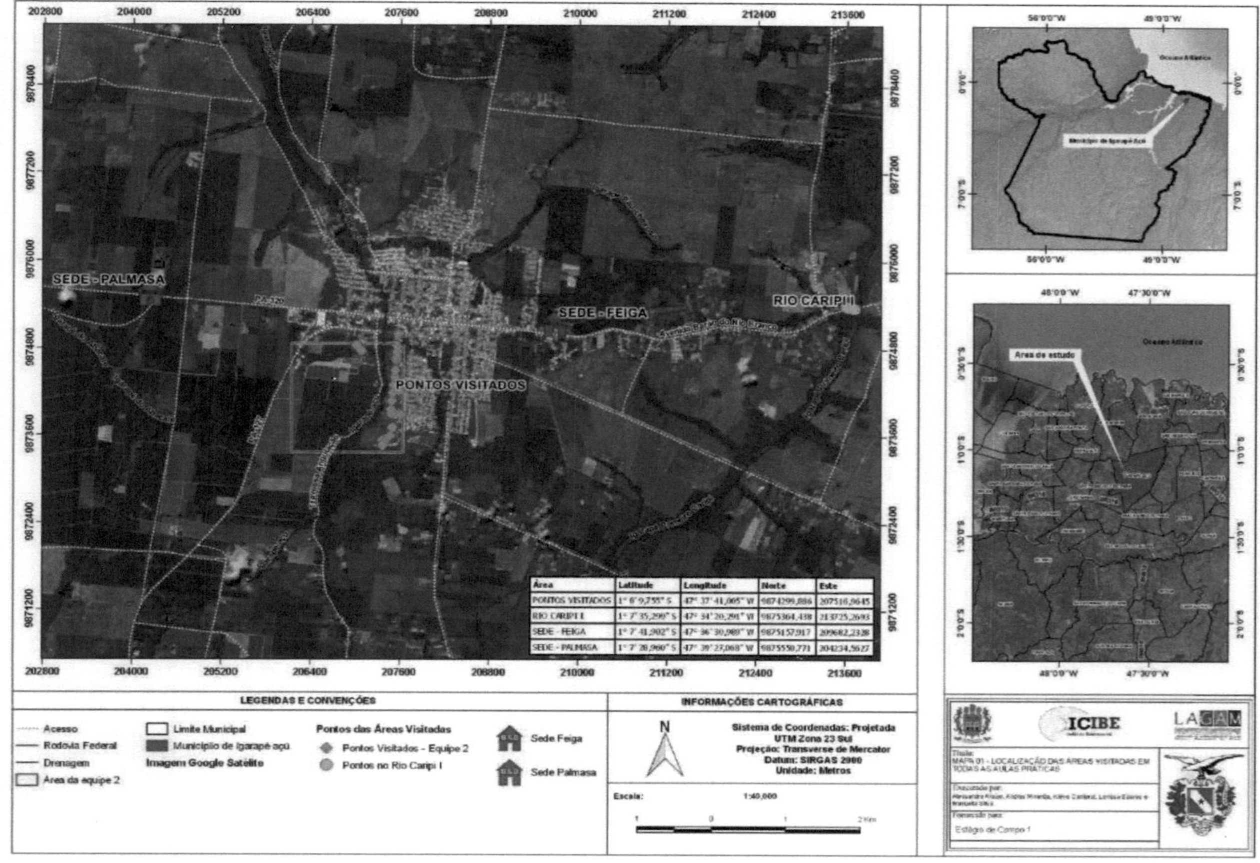

Área	Latitude	Longitude	Norte	Este
PONTOS VISITADOS	1° 6' 9,755" S	47° 37' 41,005" W	9874299,886	207516,9645
RIO CARIPI I	1° 7' 35,299" S	47° 34' 20,291" W	9875364,438	213725,2603
SEDE - FEIGA	1° 7' 41,902" S	47° 36' 30,989" W	9875157,917	209682,2338
SEDE - PALMASA	1° 7' 28,960" S	47° 39' 23,068" W	9875550,771	204234,5627

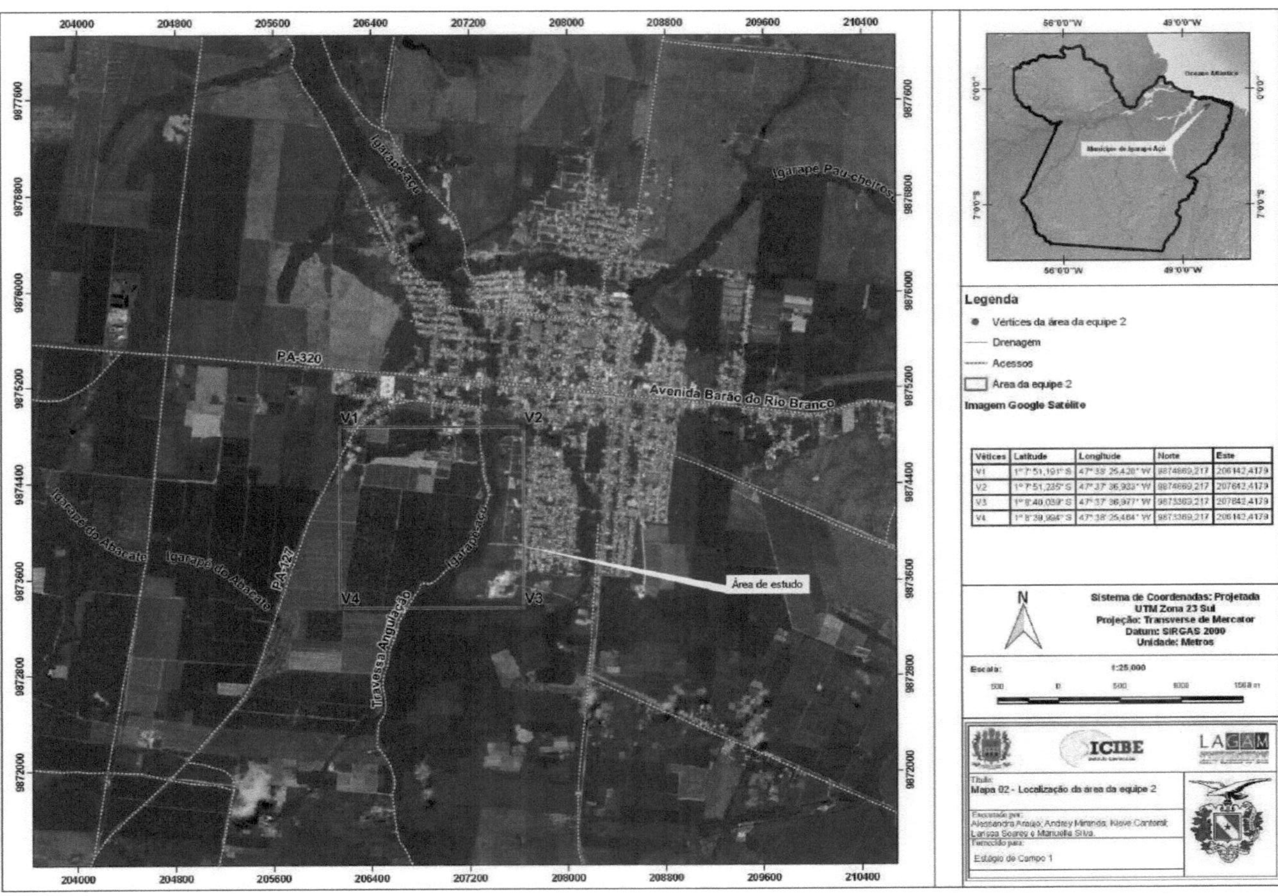

Vétices	Latitude	Longitude	Norte	Este
V1	1°7'51,191" S	47°38'25,428" W	9874869,217	206142,4179
V2	1°7'51,235" S	47°37'36,933" W	9874869,217	207642,4179
V3	1°8'40,039" S	47°37'36,977" W	9873369,217	207842,4179
V4	1°8'39,994" S	47°38'25,464" W	9873369,217	206142,4179

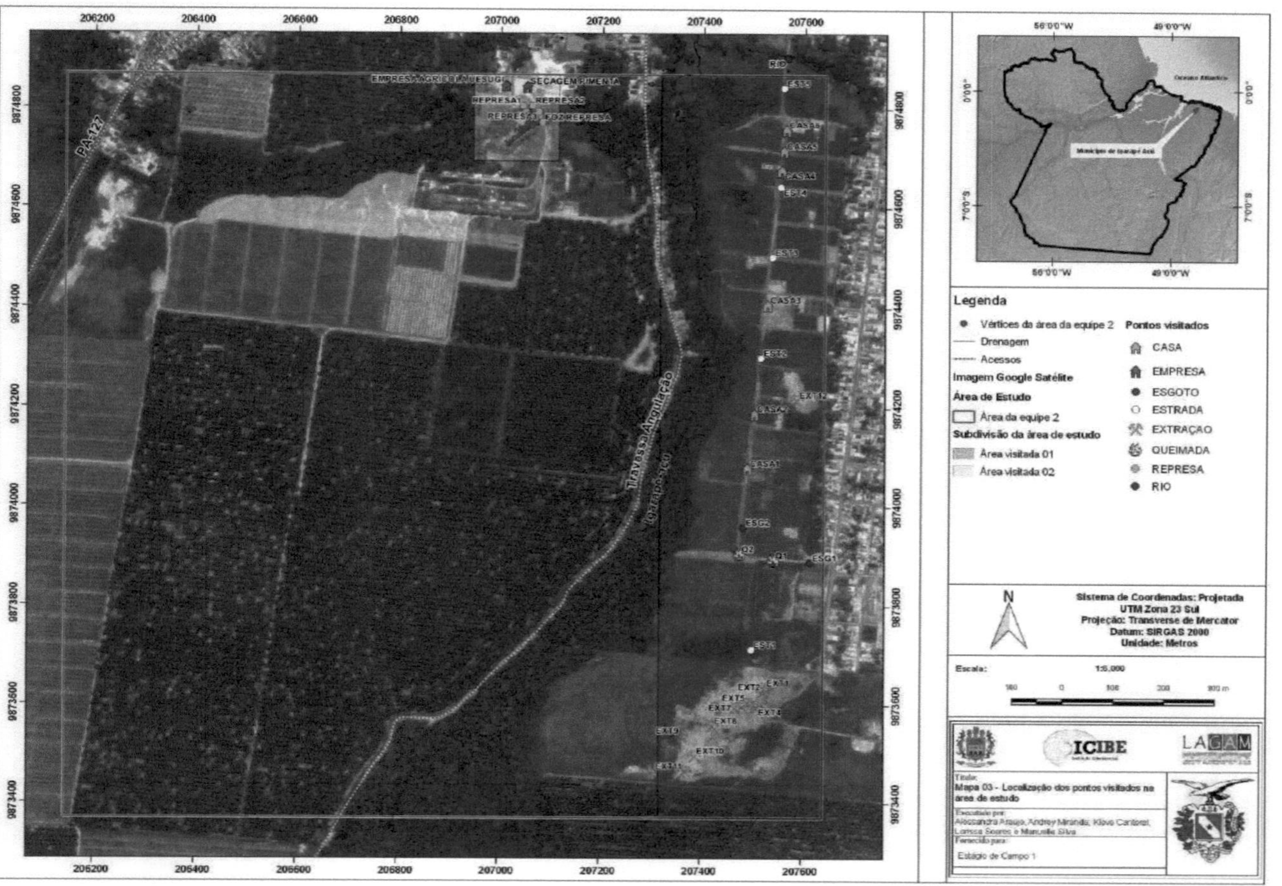

Legenda
Vértices da área da equipe 2
Drenagem
Acessos
Imagem Google Satélite
Área de Estudo
Área da equipe 2
Subdivisão da área de estudo
Área visitada 01
Área visitada 02
Pontos visitados
CASA
EMPRESA
ESGOTO
ESTRADA
EXTRAÇAO
QUEIMADA
REPRESA
RIO
Sistema de Coordenadas: Projetada
UTM Zona 23 Sul
Projeção: Transverse de Mercator
Datum: SIRGAS 2000
Unidade: Metros
Escala: 1:5.000
N
ICIBE
LAGAM
Título:
Mapa 03 - Localização dos pontos visitados na área de estudo
Executado por:
Aleccsandra Araujo, Andrey Miranda, Klévio Cantoral, Larissa Soares e Manuella Silva
Fornecido para:
Estágio de Campo 1
PA-127
Travessa Angulação
Igarapé-açu
EMPRESA AGRICOLA UESUGI
SECAGEM PIMENTA
REPRESA
Município de Igarapé Açu
Oceano Atlântico

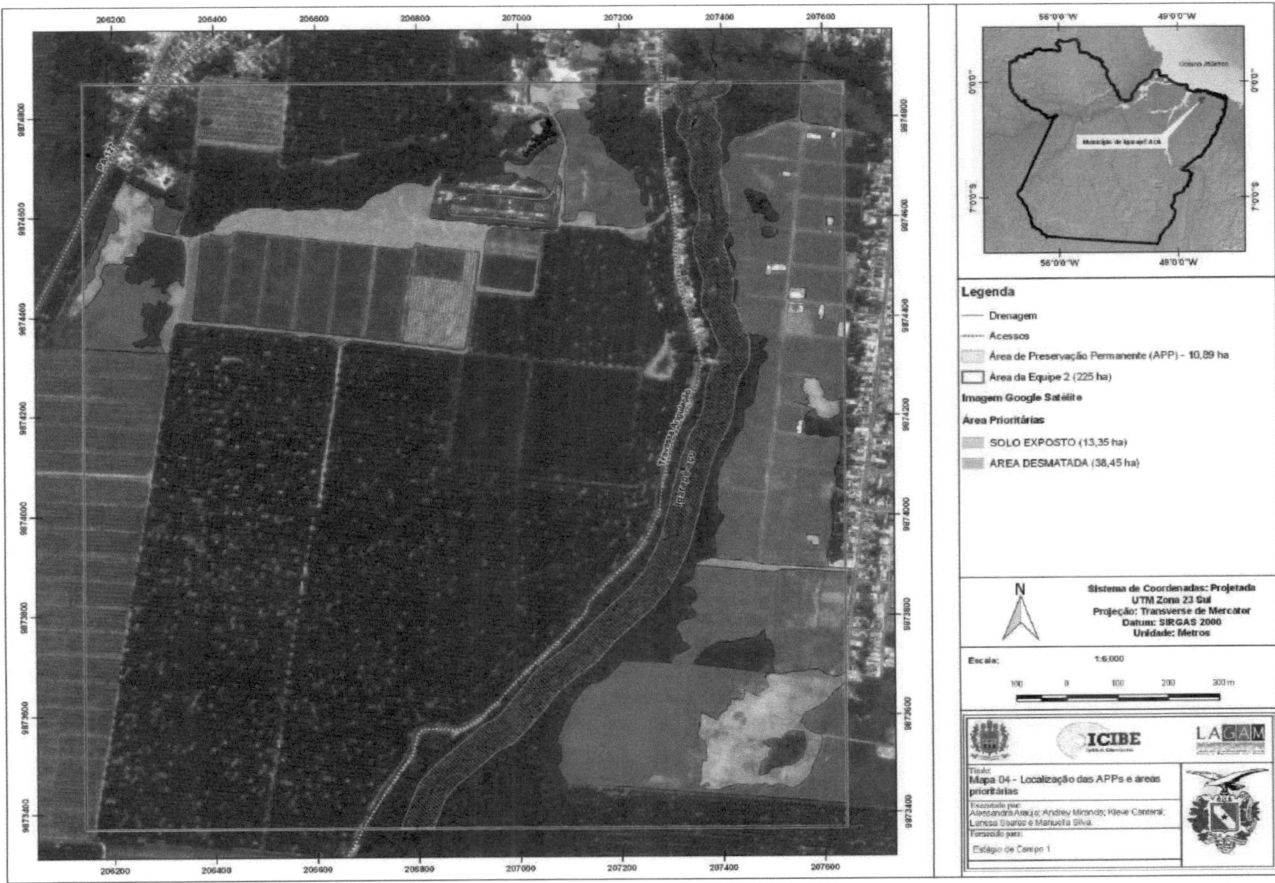

Município de Igarapé-Açu
Oceano Atlântico
Legenda
Drenagem
Acessos
Área de Preservação Permanente (APP) - 10,89 ha
Área da Equipe 2 (225 ha)
Imagem Google Satélite
Área Prioritárias
SOLO EXPOSTO (13,35 ha)
AREA DESMATADA (38,45 ha)
N
Sistema de Coordenadas: Projetada
UTM Zona 23 Sul
Projeção: Transverse de Mercator
Datum: SIRGAS 2000
Unidade: Metros
Escala: 1:6.000
ICIBE
LAGAM
Título:
Mapa 04 - Localização das APPs e áreas prioritárias
Executado por:
Alessandro Araújo; Andrev Miranda; Kleve Canteral;
Larissa Soares e Manuela Silva.
Fornecido para:
Estágio de Campo 1

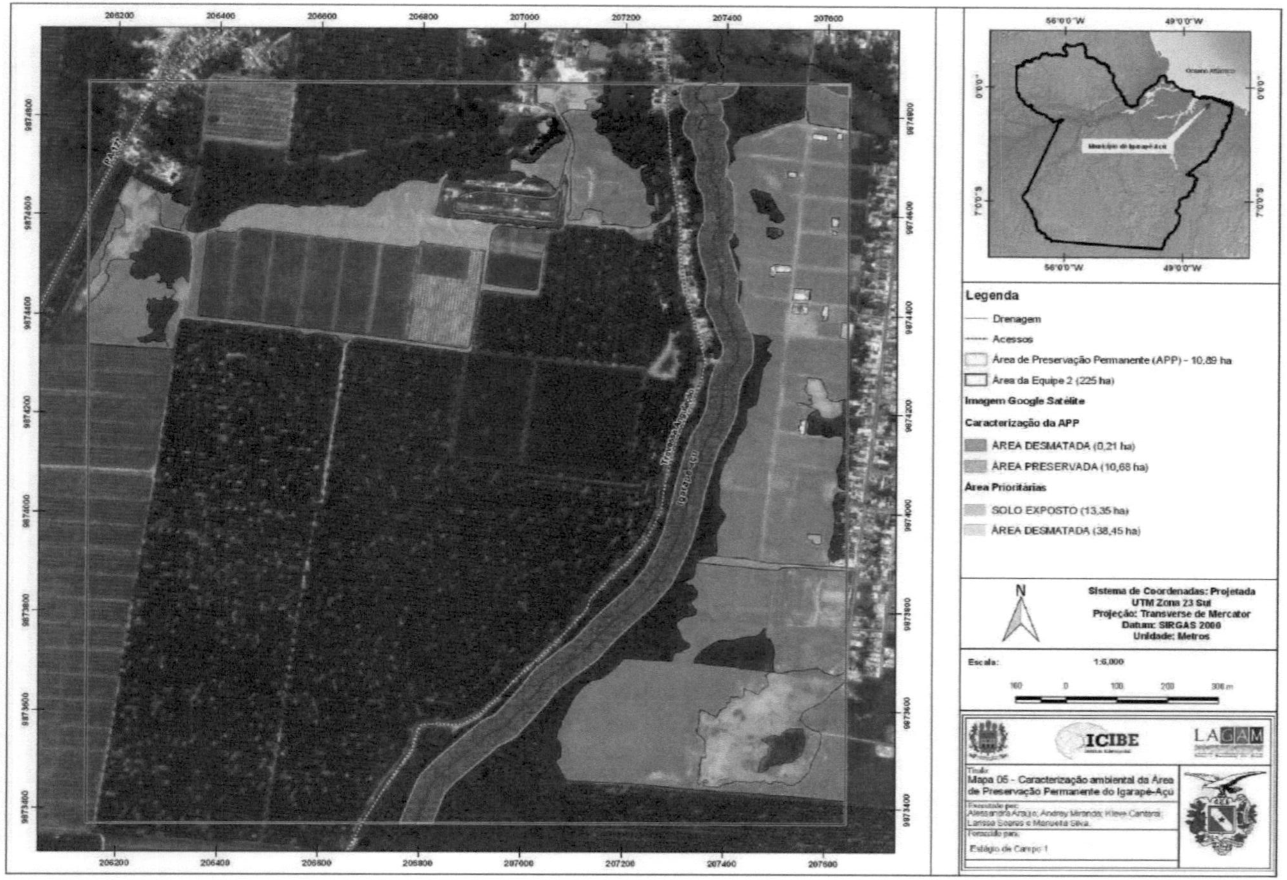

Legenda
Drenagem
Acessos
Área de Preservação Permanente (APP) - 10,89 ha
Área da Equipe 2 (225 ha)
Imagem Google Satélite
Caracterização da APP
ÁREA DESMATADA (0,21 ha)
ÁREA PRESERVADA (10,68 ha)
Área Prioritárias
SOLO EXPOSTO (13,35 ha)
ÁREA DESMATADA (38,45 ha)
Sistema de Coordenadas: Projetada
UTM Zona 23 Sul
Projeção: Transverse de Mercator
Datum: SIRGAS 2000
Unidade: Metros
Escala: 1:6.000
160 0 100 200 300 m
ICIBE
LAGAM
Título:
Mapa 05 - Caracterização ambiental da Área
de Preservação Permanente do Igarapé-Açu
Executado por:
Alessandra Araújo; Andrey Miranda; Kleve Canteral;
Larissa Soares e Manuela Silva.
Fornecido para:
Estágio de Campo 1
Município de Igarapé-Açu
Oceano Atlântico

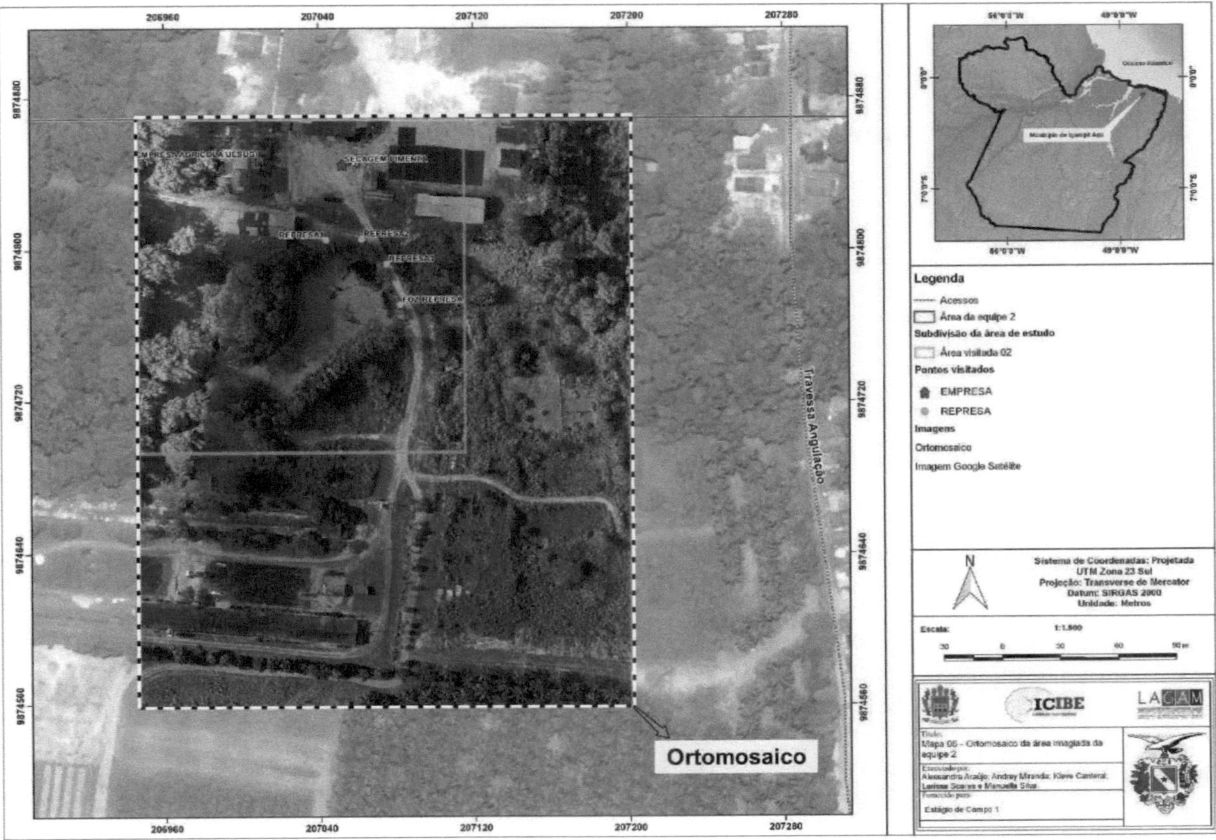

Travessa Angulação
EMPRESA AGRICOLA UESUGI
SECAGEM PIMENTA
REPRESA1
REPRESA2
REPRESA3
FOZ REPRESA
Ortomosaico
Município de Igarapé-Açu
Oceano Atlântico
Legenda
Acessos
Área da equipe 2
Subdivisão da área de estudo
Área visitada 02
Pontos visitados
EMPRESA
REPRESA
Imagens
Ortomosaico
Imagem Google Satélite
N
Sistema de Coordenadas: Projetada
UTM Zona 23 Sul
Projeção: Transverse de Mercator
Datum: SIRGAS 2000
Unidade: Metros
Escala: 1:1.500
ICIBE
LACAM
Título:
Mapa 06 - Ortomosaico da área mapeada da equipe 2
Executado por:
Alexsandra Araújo; Andrey Miranda; Kleve Canteral;
Larissa Soares e Manuella Silva
Fornecido para:
Estágio de Campo 1

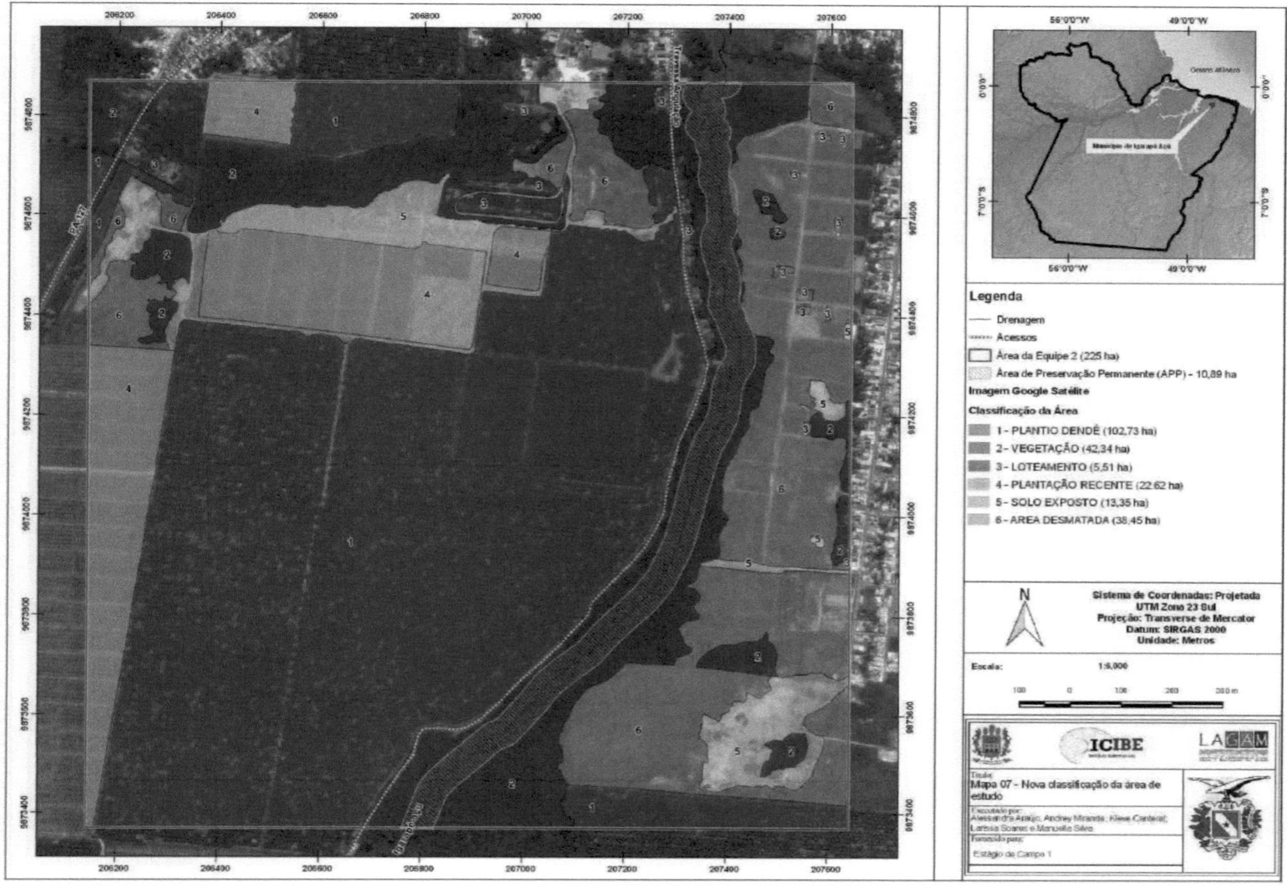

Legenda
Drenagem
Acessos
Área da Equipe 2 (225 ha)
Área de Preservação Permanente (APP) - 10,89 ha
Imagem Google Satélite
Classificação da Área
1 - PLANTIO DENDÊ (102,73 ha)
2 - VEGETAÇÃO (42,34 ha)
3 - LOTEAMENTO (5,51 ha)
4 - PLANTAÇÃO RECENTE (22,62 ha)
5 - SOLO EXPOSTO (13,35 ha)
6 - AREA DESMATADA (38,45 ha)
N
Sistema de Coordenadas: Projetada
UTM Zona 23 Sul
Projeção: Transverse de Mercator
Datum: SIRGAS 2000
Unidade: Metros
Escala: 1:6,000
ICIBE
LACAM
Título:
Mapa 07 - Nova classificação da área de estudo
Executado por:
Alessandra Araújo, Andrey Miranda, Kleve Canteral,
Larissa Soares e Manoella Silva
Fornecido para:
Estágio de Campo 1
Município de Igarapé Açu
Oceano Atlântico

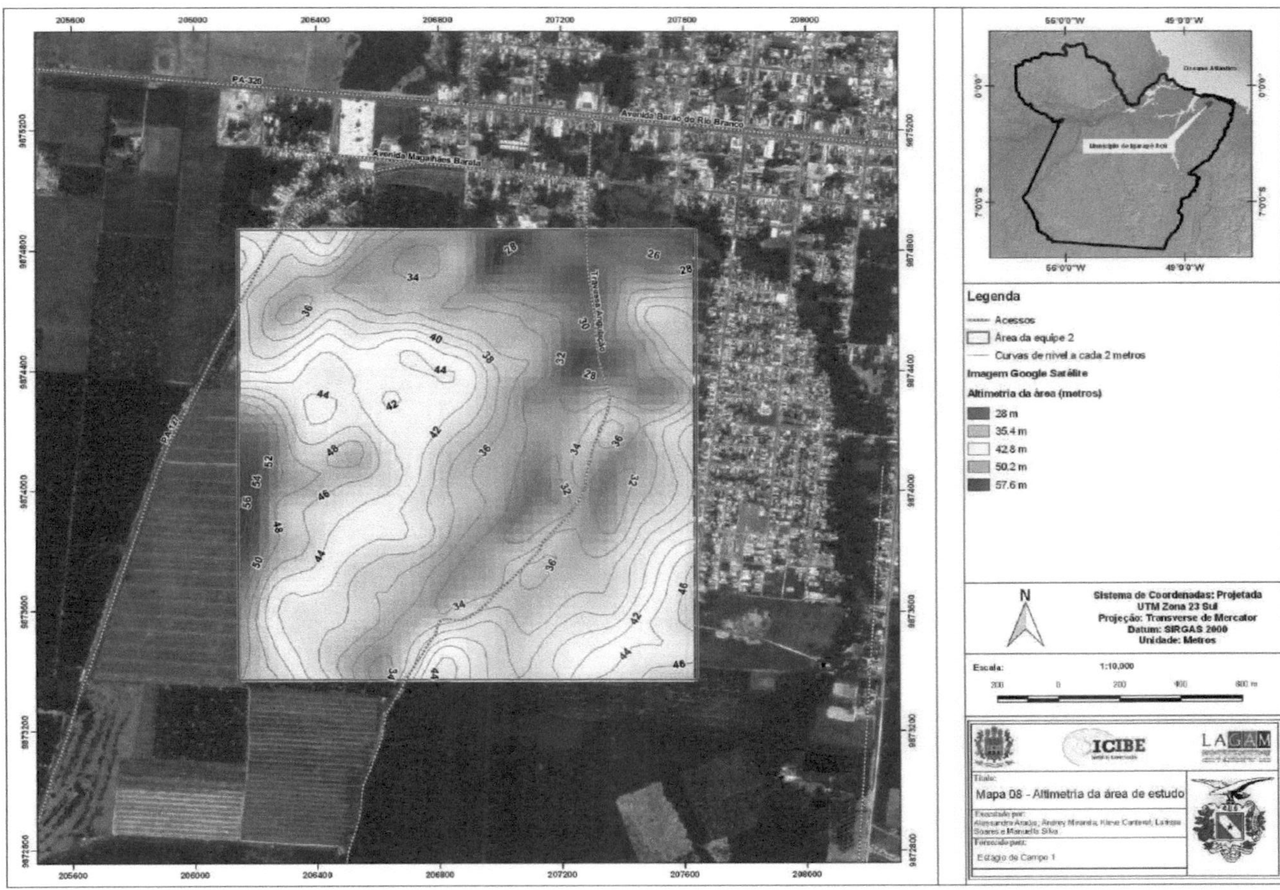
PA-320
Avenida Barão do Rio Branco
Avenida Magalhães Barata
Travessa Angelicado
PA-170
Legenda
Acessos
Área da equipe 2
Curvas de nível a cada 2 metros
Imagem Google Satélite
Altimetria da área (metros)
28 m
35.4 m
42.8 m
50.2 m
57.6 m
N
Sistema de Coordenadas: Projetada
UTM Zona 23 Sul
Projeção: Transverse de Mercator
Datum: SIRGAS 2000
Unidade: Metros
Escala:
1:10.000
ICIBE
LAGAM
Título:
Mapa 08 - Altimetria da área de estudo
Executado por:
Alexsandra Araújo; Andrey Miranda; Kleve Canteval; Larissa Soares e Manuela Silva
Torsado por:
Estágio de Campo 1

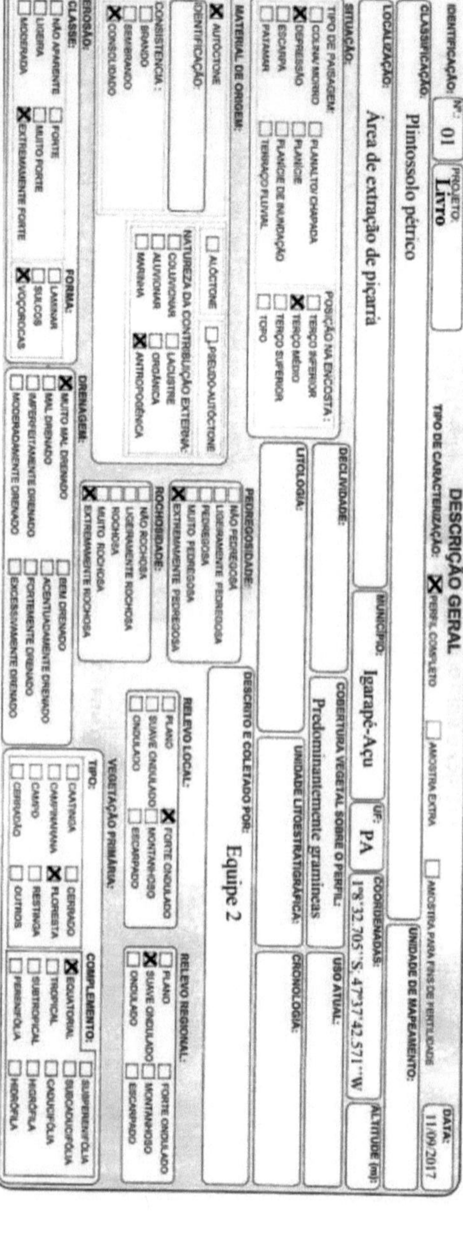

Figura 11 - Ficha para descrição morfológica dos solos no campo (continua)

Printed by Books on Demand GmbH, Norderstedt / Germany